FOOD SCIENCE AND TECHNOLOGY

MICROBIAL CONTAMINATION OF FOOD PRODUCTS

Food Science and Technology

Additional books in this series can be found on Nova's website under the Series tab.

Additional e-books in this series can be found on Nova's website under the e-book tab.

FOOD SCIENCE AND TECHNOLOGY

MICROBIAL CONTAMINATION OF FOOD PRODUCTS

GABRIEL-DĂNUȚ MOCANU
AND
OANA EMILIA CONSTANTIN
EDITORS

nova
science publishers
New York

Library of Congress Cataloging-in-Publication Data

ISBN: 978-1-53613-524-4

Published by Nova Science Publishers, Inc. † New York

CONTENTS

PREFACE

Food microbiology is the science that describes the transformations of raw materials and food products through the action of microorganisms and enzymes, and analyze the food microbiological quality and the conditions for food production to ensure food safety.

Microorganisms, due to their ability to adapt to the most diverse environmental conditions, are spread across all natural environments where they can live, multiply and evolve. In the environment and in food under certain physic-chemical storage conditions, there are two main categories of microorganisms: indigenous microorganisms, that have adapted throughout their evolution in certain microhabitats and spoilage microorganisms, that reach certain environments through contamination. Depending on the type of metabolism, three categories of microorganisms can be found in food sector: *saprophytes* able to grow, multiply, synthesize metabolites and perform biochemical transformations under uncontrolled conditions that lead to food spoilage, *opportunistic pathogens,* commonly referred to as hygienic-sanitary indicators, that are usually transmitted via feces through direct and indirect contamination, and *strict pathogens*, agents of diseases.

The microorganisms may be involved in numerous useful transformations in foods or may be acting as food spoilage or foodborne illnesses agents, with adverse effects on food quality and food safety.

Microorganisms are evolving and constantly adapt to environmental conditions. Food technology expands as well, and thus the necessity to identify the microbial contaminants of food products. This book represents a resource for food industry and academia, both focused on food safety and quality in recent years.

The *Microbial Contamination of Food Products*, consists of seven chapters. Each chapter present importance for the food microbiologists, food technologists, food processors, regulatory officials and public health workers. Furthermore, the book can be a source of valuable information on food microbiology, for the students in the field of food science and technology, food control and engineering, and management in public food.

In: Microbial Contamination of Food Product ISBN: 978-1-53613-524-4
Editors: G. Mocanu and O. Constantin

Chapter 1

MICROBIAL CONTAMINANTS OF MEAT, POULTRY AND RELATED PRODUCTS

Oana Emilia Constantin and Corina Neagu*
Department of Food Science and Engineering,
Biotechnology and Aquaculture,
Faculty of Food Science and Engineering,
Dunărea de Jos University of Galați, Galați, România

ABSTRACT

Meat is an environment in which microorganisms can quickly multiply, therefore affecting the meat quality and the safety of the consumers.

Meat, unlike other foods, is not recommended to be used in a fresh state, as it is less appreciated (the meat is less tasty and hard). The consumers expect to find unaltered meat with the right smell, flavor, taste and color, conforming to the microbiological conditions imposed by the legislation. The meat, by its composition and origin, is very sensitive to spoilage and contamination with pathogenic microorganisms, and is responsible for foodborne illnesses or poisonings.

Keywords: spoilage, meat products, foodborne, *Pseudomonas* spp.

---

* Corresponding author: cneagu@ugal.ro.

INTRODUCTION

According to European Commission Regulation No 853/2004 [1] of the European Parliament and of the Council of 29 April 2004, meat means "edible parts of the animals (domestic ungulates; poultry; lagomorphs, wild and farmed game, small and large wild game), including blood".

According to Sødring et al. [2], the definition and classification of meat varies according to the fresh meat color, the origin of animal meat, and meat flavor. Usually, the term red meat refers to meats that are red in color (bovine, porcine, ovine and caprine animals), while the term white meat is used for meats such as chicken and fish [3].

Fresh meat is meat which has not been subjected to a preservation treatment excepting chilling and freezing, vacuum or controlled atmosphere packing.

Meat preparations are produced from fresh meat, mixed with spices or/and additives, or meat that has not been processed to alter the fibrous structure of the muscles and to modify the characteristics of fresh meat.

Meat processing yields different meat products, using common meat preparations, e.g., fresh processed meat products, cured meat, dry meat, etc.

The term "meat quality" is used in different ways, depending on the training and experience of those who use it. In a wider sense, the term "meat quality" is a sum of nutritional, sensory, hygienic and technological characteristics.

To monitor the microbiological quality of the meat, the following issues are considered: raw materials, processing techniques, quality classes, preservation techniques, sensory and chemical properties.

The microbiological quality of meat is influenced by the following factors [4]:

- *chemical composition of the meat*: the meat and meat products composition consists of proteins, amino acids, vitamins, lipids (triglycerides, phosphatides); the high-water content and reduced carbohydrate content create a favorable environment for the multiplication of microorganisms, especially the spoilage bacteria;

- *pH value*: the pH of fresh meat is 5.6 - 6.8, and in cured and fermented meat products the pH is 5.5; immediately after the animal slaughter the pH value is 7 and then it descends to the value of 5, preventing the development of microorganisms;
- *the redox potential* of meat decreases after 4-6 hours after slaughter because the tissues become depleted of oxygen, ensuring the anaerobic condition in the meat depth, which promotes the growth of putrefaction bacteria;
- *water activity (a_w):* the fresh meat has an a_w of 0.98 – 0.99, and cured and processed meat has an a_w of 0.94 - 0.97; these values are optimum for microbial growth,
- *temperature:* after slaughter, the meat temperature should be reduced rapidly (-5°C to 5°C) to prevent the development of microorganisms.

The slaughtering and processing conditions are different for red meat and poultry, and the main differences are the following [5]:

- red meat carcasses are divided before evisceration, while poultry are eviscerated without splitting the carcass;
- red meat carcasses are generally skinned while poultry are processed without skin removal;
- poultry processing involves the use of abundant amounts of water as compared to red meat processing (avoiding contamination of red meat by entraining microorganisms through water from skin and intestinal content).

The sequence of these steps makes the microbiological load different for red meat and poultry meat.

For the control of microbial contaminants of meat and meat products, new measures have been developed in food safety management [6-7] (Figure 1.1).

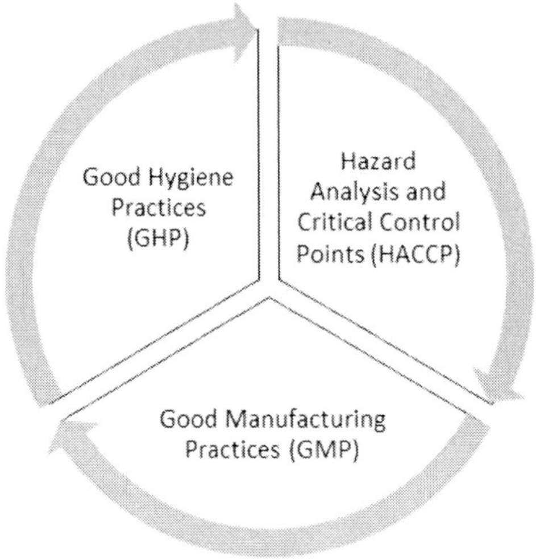

Figure 1.1. Management control systems of microbial contaminants of meat and meat products.

Spoilage Microorganisms

Meat spoilage represents the alteration of the sensory, nutritional and physico-chemical characteristics of meat under the action of microorganisms or enzymes produced by microorganisms.

Fresh meat, due to its chemical composition, can be an excellent medium for microbial growth, which can reach a high level, if allowed to continue. Microbial growth can cause the modification of meat quality, thus being unacceptable for consumers. Therefore, fresh meat is part of the most perishable food category.

In order to transform muscle into meat and to improve the sensorial and nutritive characteristics of meat and meat derived products, the complex biochemical processes that take place in the muscles after animal slaughter should be considered. Furthermore, to obtain safer meat it is critical to understand and to apply the following issues: carcass examination, preservation techniques (e.g., refrigeration), biochemical and physical changes regarding sensory characteristics [8].

The contamination sources of meat, poultry and related products may be classified into:

- endogenous contamination depending on diseased or ill animals;
- exogenous contamination depending on environmental conditions and conditions of meat processing (skinning, evisceration, splitting, toiletry, etc.)

Generally, the edible tissues of healthy animals are either sterile, or they contain a very low count of microorganisms [9]. Their deterioration begins immediately after slaughter, during the processing procedures [10] and then continues during the storage and commercialization of meat [7]. Therefore, it is desirable to identify the sources of contamination and spread of the spoilage microorganisms affecting the meat industry.

Type of Spoilage Microorganisms

Due to the chemical composition, the *indigenous and contamination microbiota* of meat consists of a great variety of microorganisms; the type and count of cells present in meat and meat products are influenced by extrinsic factors: temperature, relative humidity, oxygen concentration, radiation, mechanical and chemical factors, and intrinsic factors: chemical composition and nutrient concentration of substrate, a_w, pH, redox potential, anatomical structure [8, 11]. Meat contamination may also be influenced by the animal geographic origin and the season of the year [11].

The spoilage microflora of meat and meat products contains a few genres of interest, especially Gram-negative rod-shaped aerobic, facultative anaerobic and obligated anaerobic bacteria (Table 1.1) [12-13].

The growth, diversity and prevalence of different strains of spoilage microorganisms vary with temperature and shelf life, packaging and type of meat products [13-14]. In general, *Pseudomonas* spp. and *Shewanella* spp., are involved in the spoilage of meat, except for the meat products packaged under vacuum or in modified atmosphere.

Table 1.1. Spoilage bacteria relevant for meat products technology [12-13, 19-20]

Genus	Fresh meat	Poultry	Processed meats*
Gram positive			
Bacillus spp.	+	+	+
Brochothrix spp.	+	+	+
Clostridium spp.	+	+	++
Corynebacterium spp.	+	++	+
Staphylococcus spp.	+	+	+
Lactobacillus spp.	+	-	++
Micrococcus spp.	+	++	+
Gram negative			
Aeromonas spp.	++	+	+
Acinetobacter spp.	++	++	+
Alcaligenes spp.	+	+	+
Citrobacter spp.	+	+	-
Enterobacter spp.	+	+	+
Escherichia spp.	+	-	-
Flavobacterium spp.	+	+	-
Hafnia spp.	+	-	-
Moraxella spp.	++	+	+
Pseudomonas spp.	++	++	++
Proteus spp.	+	+	+
Psychrobacter spp.	++	+	-
Serratia spp.	+	+	-
Shewanella spp.	+	-	+

+ = known to occur; ++ = most frequently reported; - = not identified

Furthermore, meat spoilage is caused by the Enterobacteriaceae, that grow in either aerobic or anaerobic environment. Some strains belonging to *Enterobacteriaceae* originate in fecal contamination (e.g., *E. coli)*, and usually play a secondary role in the meat spoilage, because they have a minimum temperature for growth of > 7 - 8°C [15]. Psychrotrophic strains (*Hafnia alvei*, *Serratia liquefaciens* and *Enterobacter agglomerans*) are in low concentrations as compared with *Pseudomonas* spp. and *Shewanella* spp., and a competition between these bacteria is rarely found at temperatures below 7°C [15].

Table 1.2. Characterization of main spoilage bacteria of meats and meat products

Genus	Morphological and physiological characteristics	Reference
Aeromonas spp.	Gram negative/rod-shaped bacteria/non-spore forming/non-motile/facultative anaerobic/oxidase-positive/psychrophilic (pathogenic to fish) or motile mesophilic (15-38°C)	[21]
Acinetobacter spp.	Gram negative/short rod-shaped bacteria/non-spore forming/strictly aerobic/catalase-positive, oxidase-negative/optimum temperature 20-30 C	[22]
Alcaligenes spp.	Gram negative/rod-shaped bacteria or coccobacilli/motile (1 - 9 peritrichous flagella)/obligate aerobic/oxidase and catalase positive/optimal temperature 20 - 37°C	[23-24]
Bacillus spp.	Gram positive/spore forming/rod-shaped bacteria/motile (peritrichous flagella)/ obligate aerobic/mostly catalase positive /mesophilic (some species are thermophile and psychrophilic)	[25]
Brochothrix spp.	Gram positive/rod-shaped bacteria/aerobic and facultative anaerobic/ferment glucose with the production of acid/catalase positive/ oxidase negative/pH: 5.0 - 9.0 (optimum pH 7.0)/temperatures: 0 - 30°C	[26]
Citrobacter spp.	Gram negative/non-spore forming/rod-shaped bacteria/motile (peritrichous flagella)/facultative anaerobic/oxidase negative/catalase and methyl red positive/ferment glucose with acid and gas production/optimal temperature 30 - 37°C	[27]
Corynebacterium spp.	Gram positive/rod-shaped or club-shaped bacteria/aerobic or facultative anaerobic/catalase-positive/ temperature: 15 - 40°C (optimum temperature: 36°C)	[28]
Enterobacter spp.	Gram negative/non-spore forming/rod-shaped bacteria/facultative anaerobic/ ferments lactose with gas production in the presence of bile salts/oxidase- and indole-negative/optimum temperature 37 - 44°C/pH: 5.0 - 9.0 (optimum pH: 6.0 - 7.0)	[29-30]
Escherichia spp.	Gram negative/non-spore forming/rod-shaped bacteria/motile (peritrichous flagella) or non-motile/facultative anaerobe/reduces nitrates to nitrites/ferments lactose with acid and gas production/indole positive/oxidase, citrate, urease and H_2S negative/optimum temperature 37°C/pH: 4.0 - 9.0 (optimum pH: 6.0 - 7.0)	[31]

Table 1.2. (Continued)

Genus	Morphological and physiological characteristics	Reference
Flavobacterium spp.	Gram negative/rod-shaped bacteria/non-motile and motile/aerobic or facultative anaerobic/most species are psychrotolerant (4°C) and grow optimally at 20 - 30°C	[32]
Hafnia spp.	Gram negative/rod-shaped bacteria/motile (peritrichous flagella)/facultative anaerobic/methyl red, Voges-Proskauer, ornithine decarboxylase and lysine positive/arginine dihydrolase negative/pH: 4.9 - 8.25/temperature: 4 - 44°C (optimum 35°C)	[33]
Lactobacillus spp.	Gram positive/non-spore-forming rods/non-motile/catalase-negative/ferments glucose/optimal growth temperature: 30 - 40°C (thermophilic strains: 45°C)	[34]
Micrococcus spp.	Gram positive/spherical cells/strictly aerobic/catalase and oxidase positive/indole and citrate negative/optimum growth temperature: 25 - 37°C/halotolerant (5% NaCl)	[35-36]
Moraxella spp.	Gram negative/coccobacillus and bacillus shape/oxidase and catalase positive/ temperature: 20 - 42°C (optimum 35 - 37°C)	[37]
Pseudomonas spp.	Gram negative/rod-shaped bacteria/motile (polar-flagella)/temperature: 4 - 42°C (optimum 30 - 35°C)/pH: 4 - 8	[38]
Proteus sp.	Gram negative/non-spore forming/rod-shaped bacteria/motile (peritrichous flagella)/oxidatively deaminate phenylalanine/hydrolyze urea/temperature: 10 - 43°C (optimum 25°C)	[39-40]
Psychrobacter spp.	Gram negative/non-spore forming cocci or coccobacilli/osmotolerant/psychrophilic or psychrotolerant/aerobic/oxidase-positive/pH: 5.5 - 9.0 (optimum pH 6.0 - 8.0)/temperatures: -10 - 42°C/	[41-42]
Serratia spp.	Gram negative/rod-shaped bacteria/facultative anaerobic/motile rods /growth in mediums with 4% NaCl/pH: 5.0 - 9.0/temperatures: 5 - 40 C	[43]
Shewanella spp.	Gram negative/rod-shaped bacteria/motile (a single polar flagellum)/facultative anaerobic/mesophilic/produces H_2S from thiosulfate or polysulfide/incapable of fermentation	[44]

The lactic acid bacteria (LAB) which determine meat spoilage can be divided in: heterofermentative lactic acid bacteria *Lactobacillus* spp. (*L. curvatus* and *L. sakei*), *Leuconostoc* spp. [16] and *Carnobacterium* spp. [17], and homofermentative lactic acid bacteria: *Lactobacillus* spp. and *Pediococcus* spp. [17]. The homofermentative LAB only produce lactic acid, imparting an acceptable taste to meat products, whereas heterofermentative LAB produce lactic acid and CO_2, ethanol, acetic-acid, butanoic acid and acetoin, with modification of the sensory characteristics of the meat and meat products (unpleasant odor, mucus formation and color change) [17-18].

The most relevant spoilage bacteria in meat and meat products and their main physiological and morphological characteristics are shown in Table 1.1 and Table 1.2.

Yeasts and molds are not considered as the main spoilage agent in meat products [45]. Different genera of yeast and moulds isolated from meat products are shown in Table 1.3.

The yeasts' involvement in meat spoilage is associated with a high number of cells (10^7 CFU/g) and sensorial changes in meat. The more dominant yeast species are *Candida zeylanoides*, *Candida alimentaria* and *Debaryomyces hansenii* [46].

Species of *Candida*, *Cryptococcus*, *Debaryomyces* and *Yarrowia* were isolated from fresh and spoiled carcasses. *Rhodotorula* spp. and *Saccharomyces* spp. were isolated from fresh samples only, and *Trichosporon* spp. from spoiled samples only.

Fresh and Refrigerated Raw Meat

Microorganisms that cause the spoilage of red meat are divided into bacteria, molds and yeasts. Spoilage bacteria (especially proteolytic and/or putrefying bacteria) are usually responsible for the rapid degradation of protein-rich products, whereas molds and yeasts typically grow on meat products with low a_w content [7].

Table 1.3. Spoilage yeast and molds isolated from meat products [20, 45, 47-48]

Genus	Fresh meat	Poultry	Processed meats*
		Yeasts	
Candida spp.	++	+	+
Cryptococcus spp.	+	+	-
Debaryomyces spp.	+	+	++
Hansenula spp.	+	+	-
Pichia spp.	+	-	-
Rhodotorula spp.	+	+	-
Saccharomyces spp.	-	+	+
Trichosporon spp.	-	+	-
Torulopsis spp.	++	-	-
Yarrowia spp.	-	+	-
		Molds	
Alternaria spp.	+	-	+
Aspergillus spp.	+	-	++
Chrysosporium spp.	+	-	-
Cladosporium spp.	++	-	+
Fusarium spp.	+	-	+
Geotrichum spp.	++	+	+
Monilia spp.	+	-	+
Mucor spp.	++	+	+
Neurospora spp.	+	-	-
Penicillium spp.	+	+	++
Rhizopus spp.	++	+	+
Scopulariopsis spp.	-	-	+
Sporotrichum spp.	++	-	-
Thamnidium spp.	++	-	+

+ = known to occur; ++ = most frequently reported; - = not identified

The microbiology of meat at slaughter depends on the procedures applied and the meat handling, as well as the animal's state of health [49]. The initial microflora is mainly formed of mesophilic microorganisms that do not survive when the meat is cooled and stored at refrigeration temperatures. The initial number of bacteria in red meat varies between 10^2 and 10^5 CFU/cm^2 or gram [50-52]. The microbiota of the meat changes after slaughter, such that the psychotropic microorganisms will predominate during refrigeration, which represents 10% of the initial microbiota.

The predominant red meat microbiota on chilled red meat is composed of Gram negative bacteria: *Pseudomonas* spp., *Moraxella* spp., *Aeromonas* spp., *Shewanella* spp., *Psychrobacter* spp., *Acinetobacter* spp., *Enterobacteriaceae*, and Gram-positive microorganisms, including *Corynebacterium* spp., *Brevibacterium* spp., *Micrococcus* spp., *Bacillus* spp., *Lactobacillus* spp., and *Brochothrix* spp. [14, 49, 52-55].

The most common spoilage bacteria in fresh meat are *Pseudomonas* spp. and *Shewanella* spp., ubiquitous and psychrotolerant bacteria, especially *Pseudomonas fragi* and *Shewanella putrefaciens*, which represent 50 - 90% of the total microflora [15]. According to Casaburi et al. [56], the main isolated strains for *Pseudomonas* spp. were *P. fragi*, *P. fluorescens*, *P. putida*, *P. lundensis*, and *Shewanella* spp. was represented by *Shewanella putrefaciens*.

The primary raw meat microflora is mainly composed of *Pseudomonas* spp. and psychrotrophic bacteria from the *Enterobacteriaceae* family [52]; bacteria that metabolize meat components (especially carbohydrates), and can grow at pH of 5.5 - 7.0 [57]. This is the first phase of meat spoilage, which is responsible for carbohydrate catabolism with formation of short-chain fatty acids, ketones and alcohols [58]. *Pseudomonas* spp. determines the surface spoilage of meat, forming unpleasant fruity and sugary odors, and mucus on the meat surface; e.g., *Pseudomonas fragi* can be responsible for a "fruity sour" smell in meat due to the production of ethyl esters [59]. The second phase of spoilage begins when carbohydrate sources decrease in meat, and the bacteria use the amino acids as source of energy [51]. The *Pseudomonas* spp. and *Enterobacteriaceae* are responsible for the amino acid degradation. The resulting compounds are evinced by the appearance of unpleasant (putrid and sulphury) odors. The degradation of amino acids is produced by different spoilage bacteria involved in this phase, such as the following [51]:

- *Pseudomonas* spp. produce the deamination of amino acids to pyruvate, ammonia, hydrogen sulphide, methyl sulphide and dimethylsulphide.
- *Enterobacteriaceae* cause the decarboxylation of amino acids, to the final products: cadaverine, and putrescine.

Molds such as *Thamnidium* spp., *Mucor* spp. and *Rhizopus* spp. can develop on the meat surfaces and can determine defects on meat carcass surfaces stored at temperatures $<5°C$ [60].

In the case of poultry, the main spoilage microorganisms identified are *Pseudomonas* spp., followed by *Acinetobacter* spp., *Moraxella* spp. and *Psychrobacter* spp. [61]. The psychrotrophic *Enterobacteriaceae* strains such as *Enterobacter* spp. and *Serratia* spp. were also found on poultry [62]. Yeasts are also considered to play an important role in poultry spoilage, mainly the psychrotrophic *Yarrowia lipolytica* with a high proteolytic and lipolytic activity, or *Candida zeylanoides* [63].

Vacuum and Modified Atmosphere Packaging (MAP) Meat

Vacuum packaging and modified atmosphere packaging are used for the preservation of meat products. In vacuum packaging the residual oxygen is usually below 1% and can be rapidly consumed by tissue and microbial respiration, being replaced with CO_2, which can increase to 20% [51].

In MAP, in order to extend shelf life and to obtain an acceptable product, a mixture of gases is used: CO_2in combination with N_2 and/or O_2 [52]. Usually the CO_2 concentrations are variable from 10 to 40% and are responsible for the inhibition of the spoilage microflora.

The microflora of vacuum packaged and modified atmosphere packaged meat is manly characterized by psychrotrophic lactic acid bacteria [64-65], that cause sensorial changes (sourness, off-odor and discoloration) and biochemical transformations (production of gas and mucus, and pH decrease) [66-67].

The lactic acid bacteria involved in the spoilage of packed meat are mainly represented by psychrotrophic leuconostocs and lactobacilli (*Lactobacillus curvatus* and *Lactobacillus sakei*) [34]. These bacteria metabolize glucose with the production of acids (lactic, isobutanoic, isopentanoic and acetic acids), which lends meat an acid taste [51]. Björkroth et al. [68] isolated the *Leuconostoc carnosum* from ham and processed meat, as well as *Leuconostoc gasicomitatum*, also known as a spoilage-associated strain.

Garcia-Lopez et al. [51] reported that psychrotrophic *Enterobacteriaceae* and *Aeromonas* spp. determined the spoilage of vacuum-packaged and MAP beef, pork and lamb. Furthermore, two psychrotrophic species of the *Brochothrix* genus (*B. thermosphacta* and *B. campestris)* are identified as meat spoilers that can develop in MAP and vacuum-packed products of beef, pork, lamb, poultry, or fish [69-70].

Yeast species were also isolated from vacuum-packed products, such as *Kazachstania psychrophila*, from vacuum-packed beef [71] and *Yarrowia porcina* from pork meat [72].

Canned Meat Products

Canned meat can include: raw materials of animal origin (beef, pork, poultry, organs), vegetable raw materials (vegetables, different types of pasta, rice, etc.) and auxiliary materials, such as sugar, white flour, tomato paste, salt, spices, etc.

Canned products are heat-treated products, generally packaged, and the presence of spoilage microorganisms denotes the inefficiency of thermal treatments or subsequent contamination by loss of packing tightness.

The main canning microorganisms are spore-forming bacteria [73-74]:

- *mesophilic aerobic spore-forming bacteria*: *Bacillus* spp. with at optimum temperature growth of 35°C; the growth being stopped at 55°C. The endospores have a moderate resistance; they produce flat acid alteration of cans with pH > 4.6.
- *mesophilic anaerobic spore-forming bacteria: Clostridium* spp. (*Cl. sporogenes*, *Cl. bifentans*, *Cl. putrefaciens*, *Cl. hystoliticum).*

The bacteria can hydrolyze proteins, peptides, amino acids, and form hydrogen sulphide, methyl ethyl sulphides, mercaptans, ammonia, indole, etc. They grow at temperatures of 10 - 50°C and pH = 6.0 - 7.0.

- *thermophilic aerobic spore-forming bacteria:* e.g., *Bacillus stearothermophillus;* they grow in low acidity foods (pH > 4.6), and produce sugar fermentation, with formation of acids and small amounts of gas without the apparent bombardment of the can. The species only develops if preserved products are maintained at temperatures in the optimal range for thermophilic microorganisms (55-60°C).
- *sulphide anaerobic spore-forming bacteria:* e.g., *Cl. nigrificans.* The bacteria are responsible for the blackening of the product, as a result of the reaction of the solubilized iron from the metallic packaging, with the sulphur compounds formed by proteolysis. The optimum parameters for bacteria growth are: temperature 55°C and pH > 5.8.

Dry-Cured Meats

The dry-cured products are obtained by applying the following steps: adding salt and other additives to meat, maintaining the meat at low temperatures (< 5°C) until water activity (a_w) decreases to avoid the microorganisms' growth, and drying until a typical aged flavor is formed [75]. Because of the high salt concentrations and the low value of a_w, the dry-cured meat favors the growth of xerophilic fungi, such as *Aspergillus* spp., *Eurotium* spp. and *Penicillium* spp. [76-78]. Furthermore, in cellars, the microorganisms with the highest spoilage potential proved to be environmental molds. *Cladosporium* spp. have been isolated from different types of dry-cured ham: Norwegian smoked and unsmoked dry-cured meat products [79], Slovenian dry-cured meat products [78]. In addition, *Cladosporium* spp. is responsible for black spots in dry-cured fermented sausages [80].

Martín et al. [81] isolated from dry cured ham the following bacteria responsible for the product spoilage: *Serratia* spp., *Aeromonas hydrophila* and *Burholderia cepacea*. Furthermore, the *P. fluorescens* was found to be a bacterium responsible for black spots in dry-cured Iberian hams [82].

Fermented Meat and Poultry

The initial microbiota of fermented meat and poultry contains *Pseudomonas* spp. and *Enterobacteriaceae* strains, but the evolution of these bacteria differs according to the type of product obtained, the technological process and the origin country of the product. For example, in French products, the concentration of these bacteria remained stable during the fermentation process, but their number was significantly reduced in the final product, after the maturation phase, whereas in Greek products the bacteria were gradually eliminated [83].

Biochemistry of Spoilage

The main mechanisms of meat spoilage are: microbial and autolytic enzymatic spoilage, and lipid oxidation [84]. Important physicochemical changes occur in the aqueous phase of the meat which contains amino acids, oligopeptides and carbohydrates [85]. The substrates used by spoilage bacteria are diverse (Table 1.4), although the substrate initially catabolized by aerobic and anaerobic bacteria is glucose, and its reduction can be seen before the occurrence of any organoleptic spoilage manifestation. When glucose is exhausted from the environment, meat spoilage bacteria use other available substrates, such as glucose-6-phospate, lactic acid, lactate, gluconate, gluconate 6-p, pyruvate, propionate, formate, ethanol, acetate, creatine, creatinine, citrate, amino acids, urea and water-soluble proteins [85-86].

The main end-products formed by aerobic and anaerobic spoilage bacteria are: sulphur compounds, esters, ketones, alcohols, aromatic hydrocarbons, biogenic amines [85], and the presence of these compounds can be visible or not on meat products (Table 1.5.).

Table 1.4. Substrates used by major meat spoilage bacteria [5, 85, 87]

Substrates	Microorganism involved	
	Aerobic condition	**Anaerobic condition**
Glucose	*Pseudomonas* spp., *Shewanella putrefaciens*, *Brochothrix thermosphacta*, *Enterobacter* spp.	*Lactobacillus* spp., *Carnobacterium* spp., *Leuconostoc* spp., *Clostridium esterthetheticum*, *Shewanella putrefaciens*, *Brochothrix thermosphacta*, *Enterobacter* spp.
Amino acids	*Pseudomonas* spp., *Moraxella* spp., *Shewanella putrefaciens*, *Acinetobacter* spp., *Psychrobacter* spp., *Enterobacter* spp.	*Shewanella putrefaciens*, *Enterobacter* spp., *Lactobacillus* spp., *Clostridium esterthetheticum*
Lactic acid	*Pseudomonas* spp., *Shewanella putrefaciens*, *Acinetobacter* spp., *Psychrobacter* spp., *Enterobacter* spp.	*Lactobacillus* spp.

Pathogenic Microorganisms in Meat

Fresh and frozen meat and related meat products can represent important sources of foodborne pathogens. The most common pathogenic bacteria found in the meat industry are *Salmonella* spp., *Campylobacter jejuni/coli*, *Yersinia enterocolitica*, Verotoxigenic *E. coli*, as well as *Listeria monocytogenes*, *Clostridium* spp. and *Staphylococcus aureus* [93]. According to Mor-Mur and Yuste [94], the outbreaks determined by *chicken consumption* are usually caused by products contamination with the fallowing pathogens: *Salmonella* spp., *Staphylococcus aureus*, *Bacillus cereus*, *Aeromonas hydrophila*, *Listeria monocytogenes* and *Yersinia enterocolitica.*

The growth conditions of pathogen bacteria capable to contaminate meat and meat products are shown in Table 1.6.

The most important pathogenic microorganisms with incidence in the meat industry will be described below.

Table 1.5. Meat main defects caused by spoilage microorganisms

Effect	Microorganism involved	Meat source	Reference
	Bacteria		
Slime	*Acinetobacter*, *Alcaligenes*, *Bacillus*, *Leuconostoc*, *Moraxella*, *Micrococcus*, *Pseudomonas*, *Streptococcus*	raw meat	[60]
Slime or discoloration	*Micrococcus*, *Acinetobacter*	bacon refrigerated	[88]
Slime and pigmentation (purple)	*Janthinobacterium lividum*	roast beef	[51]
Pigmentation (spots)	*Serratia marcescens* (red), *Pseudomonas syncyanea (*blue), *Flavobacterium (*yellow), *Chromobacterium lividum* (green-blue or black)	raw meat	[60]
Putrid odor and gas (H_2 and CO_2) - Blown pack	*Clostridium esterthetic*um	vacuum-packaged meat	[89]
Buttery off-odors	*Lactococcus piscium*	vacuum-packaged refrigerated beef	[90]
	Molds		
white spots	*Acremonium* spp., *Chrysosporium pannorum*, *Sporotrichum carnis*	raw meat	[60, 91]
black spot	*Aureobasidium pullulans*, *Cladosporium cladosporioides*, *Cladosporium herbarum*, *Penicillium hirsutum*,	raw meat	[92]
blue-green spots	*Penicillium corylophilum*, *Penicillium expansum*, *Penicillium asperulum*	raw meat	[60, 91]
"whisker" spots	*Mucor racemosus*, *Thamnidium elegans*	raw meat	[60]

Salmonella spp.

Salmonella spp. are Gram-negative bacteria from the *Enterobacteriaceae* family. More than 2500 serotypes of the genus

Salmonella are known to scientists, based on the differentiated somatic, flagellar and capsular antigens in the Kauffman White scheme [98].

Nowadays, two species of *Salmonella* are recognized: *S. enterica* and S. *bongori* [95]. *S. enterica* consists of six subspecies; the serovars responsible for foodborne disease are part of subspecies I (*S. enterica* subsp. *enterica*) [99]. The different serotypes of *Salmonella enterica* can be divided into two categories: non-typhoid and typhoid [97]. *S. enterica* serotypes which are important for the food industry are: serotypes *Typhi*, *Paratyphi A*, and *Paratyphi B* (associated with humans) and serotypes *Enteritidis* (broilers and hens) and T*yphimurium* (associated with pigs, cattle and chickens) [100].

Table 1.6. Bacterial pathogen growth conditions [93, 95-97]

Pathogen	Oxygen requirement	pH (pH optimum)	Temperature, °C (temperature optimum)
Aeromonas hydrophila	facultative anaerobic	5.5 - 10	5 - 45°C (28°C)
Bacillus cereus	aerobic	4.9 - 9.3 (6.8 - 7.2)	4 - 48°C (28 - 35°C)
Campylobacter jejuni / *Campylobacter coli*	microaerophilic (3 - 5% O_2/2–10% CO_2)	4.9 - 9.5 (6.5–7.5)	30 - 45°C (42°C)
Clostridium botulinum	anaerobic	4.6 - 9 (6 - 7.5)	10 - 48°C (43 - 45°C)
Clostridium perfringens	anaerobic	5 - 9	10 - 52°C
VCEC Escherichia coli	facultative anaerobic	4 -10	6 - 49°C (37 °C)
Listeria monocytogenes	facultative anaerobic	4.4 - 9.4 (7.0)	0 - 45°C (37 °C)
Salmonella spp.	facultative anaerobic	3.7 - 9.5	5 - 46°C
Staphylococcus aureus	facultative anaerobic	4 - 10 (6 - 7)	7 - 50°C (37 °C)
Yersinia enterocolitica	facultative anaerobic	4.2 - 9.4 (7.2 - 7.4)	-2 - 43°C (25 - 29 °C)

The natural habitat of most *Salmonella* strains is the gastrointestinal tract of animals, and environmental places (water, food, animal feed, etc.) as a result of fecal contamination [101].

Salmonella spp. shows sensitivity to salts, being capable to grow in foods products with 2% NaCl at an optimum temperature of growth [95], and grows at a_w - 0.93 in foods (optimum of 0.995) [99]. The *Salmonella* spp. eventually present on meat and meat products is destroyed by pasteurization (approx. 70°C) and can survive in freezing conditions [93].

The non-typhoid *Salmonella* spp. is capable to produce salmonellosis, which according to WHO [102] is a disease characterized by fever, abdominal pain, diarrhea, nausea and vomiting. The infection occurs at the ingestion of a concentration of 10^5 - 10^6 cells. The typhoid *Salmonella* spp. (*S. typhi* and *S. paratyphi*) are responsible for enteric fever.

Campylobacter spp.

Campylobacter spp. are Gram-negative, spiral rods bacteria belonging to the Campylobacteriaceae family; they are reported to produce some toxins [97]. The species pathogenic to humans are *Campylobacter jejuni*, *Campylobacter coli*, *Campylobacter lari* and *Campylobacter upsaliensis* [103].

The natural habitats of *Campylobacter* spp. are: the intestinal tract of wild animals, domestic animals (cattle, sheep, pigs), birds (chickens) and the water supply as a result of fecal contamination [97, 104]. *Campylobacter* spp. were frequently isolated from fresh beef, pork and poultry. In thermally processed foods, their presence is associated with contamination after thermal treatment, or if thermal treatment has been improperly applied.

Campylobacter spp. determine campylobacteriosis in humans at the ingestion of a concentration of 10^2 cells. The symptoms of campylobacteriosis are diarrhea, accompanied or not by nausea, vomiting, abdominal pain, fever, chills, alteration of the general condition [105].

Yersinia spp.

Yersinia spp. are Gram-negative, rod-shaped bacteria belonging to the *Enterobacteriaceae* family. *Yersinia* spp. are psychrotrophic, being capable of growing at refrigeration temperatures.

The natural habitats of *Yersinia* spp. are the intestinal microbiota of domestic and wild birds, and the human microbiota. The meat products (pork, beef, lamb, etc.), can be contaminated with *Yersinia*, when the sanitary practices and the thermal treatment have been improperly applied.

The foodborne species responsible for yersiniosis is reported to be *Yersinia enterocolitica* [97]. Yersiniosis in humans is characterized by the following symptoms: fever, abdominal pain and diarrhea at the ingestion of a concentration of 10^7 cells.

Listeria spp.

Listeria spp. are Gram-positive, motile, rod-shaped organisms belonging to the Listeriaceae family. The *Listeria monocytogenes* is the most important species, being pathogenic to humans and animals [105].

Listeria monocytogenes is commonly found in natural environments (soil, water, vegetable residue, animal feces), and has been isolated in the intestinal content of poultry.

The *L. monocytogenes* can be present on meat and meat products as an accidental contaminant and is destroyed by pasteurization (approx. 75°C); it can multiply in meats preserved by refrigeration [93]. *L. monocytogenes* is a post-processing contaminant mainly during additional processes such as peeling, slicing, and repackaging [106].

Listeriosis is produced by *L. monocytogenes* and shows the following symptoms: fever, muscle aches, nausea, vomiting and gastric pain, but some more severe symptoms may present such as: head and cervical stiffness, convulsions, loss of balance.

E. coli

E. coli are Gram-negative, non-spore forming rods bacteria, from the Enterobacteriaceae family.

Pathogenic *E. coli* associated with food are divided as following [105, 107-108]:

1. enteropathogenic *E. coli* (EPEC);
2. verocytotoxigenic *E. coli* (VCEC)/*Shiga* toxin producing *E. coli* (STEC) and enterohemorrhagic *E. coli* (EHEC);
3. enterotoxigenic *E. coli* (ETEC);
4. enteroaggregative *E. coli* (EAEC);
5. enteroinvasive *E. coli* (EIEC);
6. diffuse adherent *E. coli* (DAEC).

The natural habitats of *E. coli* spp. are the gastrointestinal tract of animals (cows, sheep, goats, etc.), humans, etc. The fresh meat and meat products can be contaminated with *E coli* [108-109]:

- from animals as the primary source, from the feces or the gastrointestinal tract during slaughter;
- from the workers when the hygiene has been improperly applied, from contaminated water;
- during the different steps of the food chain, e.g., during meat slicing, or packaging, food handling, etc.

The common symptoms of disease associated with pathogenic E. coli (EPEC, VCEC or STEC, EHEC, ETEC, EAEC, EIEC): fever, diarrhea, abdominal pain, nausea, vomiting [105, 107-108]:

Clostridium spp.

Clostridium spp. are Gram-positive, rod-shaped spore-forming bacteria. The majority are saprophytes, but four species are human pathogens (*Cl. perfringens*, *Cl. botulinum*, *Cl difficile* and *Cl. tetani*) [105]. From the species above, *Cl. perfringens* and *Cl. Botulinum* are mainly involved in fresh meat contamination.

Clostridium botulinum is commonly found in natural environments: soil, water, sludge, waste water and feces. The bacterium produces the botulinum toxin, when the environment is characterized by low-oxygen concentrations [110], being responsible for foodborne poisoning. It can be found in canned, thermally treated or fermented meat products.

Clostridium perfringens is commonly found in primary sources (soil, dust, feces), and secondary sources (all raw and auxiliary materials that come in contact with primary sources), being responsible for food poisoning. It can be found in meats and meat products contaminated through the addition of additives, and during post-processing [111].

Staphylococcus aureus

Staphylococcus aureus is a Gram positive, non-motile, non-spore bacteria, from the Staphylococcaceae family.

This bacterium has the ability to grow under conditions that for other microorganisms are restrictive: $a_w = 0.86$ and in media with 15% NaCl [93, 97].

Naturally, *Staphylococcus aureus* is found in the buccal/nasal microbial mucosa, in the skin, at the hair base or feathers, in purulent wounds, etc.

The meat and meat products can be contaminated with *Staphylococcus aureus* under the following circumstances [112]:

- inadequate storage of food;
- insufficient cooking and keeping food at temperatures that exceed the danger zone (5 - 63°C);
- poor personal hygiene of the workers handling the raw materials.
- improper sanitation of production/storage areas of food.

S. aureus is responsible for foodborne poisoning, characterized by the following symptoms: abdominal cramps, nausea, vomiting and diarrhea.

EU Regulations

According to the European Regulations [113] regarding microbiological criteria for foodstuffs, meat and meat products must fulfill the following microbiological conditions (Table 1.7).

Table 1.7. Food safety and process hygiene criteria for meat and meat products

Food category	Microorganisms	Limits
	Food safety criteria	
Red minced meat	*Salmonella*	Absence in 10 g
Poultry minced meat	*Salmonella*	Absence in 10 g
Meat products	*Salmonella*	Absence in 25 g
Poultry meat products	*Salmonella*	Absence in 25 g
Gelatin and collagen	*Salmonella*	Absence in 25 g
	Process hygiene criteria	
Minced meat/Mechanically separated meat	aerobic colony count	$5x10^5$ - $5x10^6$ CFU/g
	E. coli	50 - 500 CFU/g
Meat preparations	*E. coli*	500 - 5000 CFU/g

*CFU-colony forming units

Conclusion

Meat is a complex and nutritionally valuable food, due to its carbon and nitrogen sources, vitamins and high free water content (67 - 71%); therefore, it can ensure the growth of microorganisms in general, and bacteria, in particular. For this reason, it is very important to identify meat contamination sources and to apply the adequate measures to control the pathogenic bacterial contamination.

References

[1] European Commission (EC). (2004). "Regulation (EC) No 853/2004 of the European Parliament and of the Council of 29 April 2004 Laying Down Specific Hygiene Rules for Food of Animal Origin". *Official Journal of the European Union*, L *139*, 30 April 2004. http://data.europa.eu/eli/reg/2004/853/oj Accessed September 2, 2017 (EC 2004, 32-33).

[2] Sødring, M., Oostindjer, M., Dragsted, L. O., Haug, A., Paulsen, J. E. & Egelandsdal, B. (2017). "Meat and Cancer Evidence for and Against". In *New Aspects of Meat Quality from Genes to Ethics*, edited by Peter P. Purslow, 479-499. Woodhead Publishing, Cambridge (Sødring et al. 2017, 479).

[3] Oostindjer, Marije., Jan, Alexander., Gro, V. Amdam., Grethe, Andersen., Nathan, S. Bryan., Duan, Chen., Denis, E. Corpet., Stefaan, De Smet., Lars, Ove Dragsted., Anna, Haug., Anders, H. Karlsson., Gijs, Kleter., Theo, M. de Kok., Bård, Kulseng., Andrew, L. Milkowski., Roy, J. Martin., Anne-Maria, Pajari., Jan, Erik Paulsen., Jana, Pickova., Knut, Rudi., Marianne, Sødring., Douglas, L. Weed. & Bjørg, Egelandsdal. (2014). "The Role of Red and Processed Meat in Colorectal Cancer Development: A Perspective". *Meat Science*, *97*(4), 583-596. doi: 10.1016/j.meatsci.2014.02.011 (Oostindjer et al. 2014, 583).

[4] Deak, Tibor. (2007). "Yeasts in Specific Types of Foods". In *Handbook of Food Spoilage Yeasts*, edited by Tibor Deak, 117-220. CRC press, London (Deak 2007, 168).

[5] Corry, J. E. L. (2007). "Spoilage Organisms of Red Meat and Poultry". In *Microbiological Analysis of Red Meat, Poultry and Eggs*, edited by G. C. Mead, 101-122. Woodhead Publishing Limited, Cambridge, England (Corry 2007, 101, 110-111).

[6] Koutsoumanis, K. P., Geornaras, I. & Sofos, J. N. (2006). "Microbiology of Land Muscle Foods". In *Handbook of Food Science, Technology and Engineering*, vol. 1, edited by Y. H. Hui, 52.1–52.43. CRC Press, Boca Raton, FL (Koutsoumanis, et al. 2006, 52.2).

[7] Zhang, Wangang. & Yimin, Zhang. (2017). "Spoilage Microorganisms in Meat Products". In *Food Spoilage Microorganisms: Ecology and Control*, edited by Yanbo Wang, Wangang Zhang, and Linglin Fu, 77-99, CRC Press, Boca Raton, FL (Zhang and Zhang 2017, 78-79, 85).

[8] Cruz-Monterrosa, Rosy. & Isabel, Guerrero-Legarreta. (2012). "Postmortem Handling". In *Handbook of Meat and Meat Processing* (2nd Edition), edited by Y. H. Hui, 315-321. CRC Press, Boca Raton, FL (Cruz-Monterrosa and Guerrero-Legarreta 2012, 316).

[9] Martin, J. M. (2007). "Beef Quality and Tainting". In *Handbook of Meat, Poultry and Seafood Quality*, edited by Leo M. L. Nollet, 327-331. Blackwell Publishing (Martin 2007, 328).

[10] Sofos, J. N. (1994). "Microbial Growth and its Control in Meat, Poultry and Fish". In *Advances in Meat Research: Quality Attributes and their Measurement in Meat, Poultry and Fish Products*, edited by A. M. Pearson, and T.R. Dutson, 353-403. Glasgow: Chapman and Hall (Sofos 1994, 353).

[11] Paramithiotis, Spiros., Skandamis, P. N., George-John, E. & Nychas. (2009). "Insights into Fresh Meat Spoilage". In *Safety of Meat and Processed Meat*, edited by Fidel Toldrá, 55-82. Springer, New York (Paramithiotis et al. 2009, 55).

[12] Comi, Giuseppe. & Lucilla, Iacumin. (2012). "Meat and Fish Products". In *Food Microbiology*, edited by Marco Gobetti, Massimo Vincenzini, Giovanni A. Farris, Erasmo Neviani, 153-196. Ambrosiana, Milano, Italy (Comi 2012, 153-171).

[13] Doulgeraki, Agapi I., Danilo, Ercolini., Francesco, Villani. & George-John, E. Nychas. (2012). "Spoilage Microbiota to the Storage of Raw Meat in Different Conditions". *International Journal of Food Microbiology*, *157*, 130-141. doi: 10.1016/j.ijfoodmicro.2012.05.020 (Doulgeraki 2012, 130, 133-134).

[14] Ercolini, Danilo., Ilario, Ferrocino., Antonella, Nasi., Maurice, Ndagijimana., Pamela, Vernocchi., Antonietta, La Storia., Luca, Laghi., Gianluigi, Mauriello., Elisabetta, M. Guerzoni. & Francesco, Villani. (2011). "Monitoring of Microbial Metabolites and Bacterial Diversity in Beef Stored in Different Packaging Conditions". *Applied and Environmental Microbiology*, *77*, 7372–7381. doi:10.1128/AEM. 05521-11 (Ercolini 2011, 7374, 7380).

[15] Comi, Giuseppe. (2017). "Spoilage of Meat and Fish". In *The Microbiological Quality of Food - Foodborne Spoilers*, edited by Antonio Bevilacqua, Maria R. Corbo, and Milena Sinigaglia, 179-210. Elsevier Ltd (Comi 2017, 184, 186).

[16] Hu, Ping., Guanghong, Zhou Xinglian Xu., Chunbao, Li. & Yanqing, Han. (2009). "Characterization of the Predominant Spoilage Bacteria in Sliced Vacuum-Packed Cooked Ham Based on 16S rDNA-DGGE". *Food Control*, *20*, 99-104. doi: 10.1016/j.foodcont. 2008.02.007 (Hu et al. 2009, 102).

[17] Iulietto, Maria F., Paola, Sechi., Elena, Borgogni. & Beniamino, T. Cenci-Goga. (2015). "Meat Spoilage: A Critical Review of a Neglected Alteration Due to Ropy Slime Producing Bacteria". *Italian Journal of Animal Science*, *14*(3), 4011. doi: 10.4081/ijas.2015.4011 (Iulietto et al. 2015, 320).

[18] Krockel, L. (2013). "The Role of Lactic Acid Bacteria in Safety and Flavour Development of Meat and Meat Products". In *Lactic Acid Bacteria - R and D for Food, Health and Livestock Purposes*, edited by M. Kongo, 129-152, InTech Publ. (Krockel 2013, 132).

[19] Fernandes, Rhea. (2009). "Chilled and Frozen Raw Meat, Poultry and their Products". In *Microbiology Handbook*, edited by Rhea Fernandes, 1-52. Leatherhead Food International Ltd. (Fernandes 2009, 6-7).

[20] Jay, J. M. (2005). "Processed Meats and Seafoods". In *Modern food microbiology*, 7th edition, edited by J.M. Jay, 101-124. Springer Science and Business Media (Jay 2005, 102).

[21] Percival, S. L. & Williams, D. (2013). "*Aeromonas*". In *Microbiology of Waterborne Diseases: Microbiological Aspects and Risks*, edited by S.L. Percival, M.V. Yates, D. Williams, R. Chalmers, and N. Gray, 49-64. Academic press, London, UK (Percival and Williams 2013, 49).

[22] Percival, S. L. & Williams, D. (2013). "*Acinetobacter*". In *Microbiology of Waterborne Diseases: Microbiological Aspects and Risks*, edited by S.L. Percival, M.V. Yates, D. Williams, R. Chalmers, and N. Gray, 35-48. Academic press, London, UK (Percival and Williams 2013, 35).

[23] Austin, Brian. (2014). "The Family *Alcaligenaceae*". In: *The Prokaryotes. Alphaproteobacteria and Betaproteobacteria* (4rd edition), edited by Eugene Rosenberg, Edward F. DeLong, Stephen Lory, Erko Stackebrandt, and Fabiano Thompson, 729-738. Springer, Berlin, Heidelberg (Austin 2014, 733-39).

[24] Busse, Hans J. & Andreas, Stolz. (2006). "*Achromobacter, Alcaligenes* and Related Genera". In *The Prokaryotes - A Handbook on the Biology of Bacteria*, vol. *5* (3rd edition), edited by Martin Dworkin, Stanley Falkow, Eugene Rosenberg, Karl-Heinz Schleifer, and Erko Stackebrandt, 675-700. Springer New York (Busse and Stolz 2006, 675-78).

[25] Logan, N. A. & Rodríguez-Díaz, M. (2006). "*Bacillus* spp. and Related Genera". In *Principles and Practice of Clinical Bacteriology*, 2nd, edited by Stephen H. Gillespie, Peter M. Hawke, 139-158, John Wiley & Sons (Logan and Rodríguez-Díaz 2006, 140).

[26] Stackebrandt, Erko. & Dorothy, Jones. (2006). "The genus *Brochothrix*". In *The Prokaryotes - A Handbook on the Biology of*

Bacteria, vol. *4* (3rd edition), edited by Martin Dworkin, Stanley Falkow, Eugene Rosenberg, Karl-Heinz Schleifer, and Erko Stackebrandt, 477-491, Springer, New York (Stackebrandt and Jones 2006, 678-681).

[27] Borenshtein, Diana. & David, B. Schauer. (2006). "The genus *Citrobacter*". In *The Prokaryotes - A Handbook on the Biology of Bacteria*, vol. *6* (3rd edition), edited by Martin Dworkin, Stanley Falkow, Eugene Rosenberg, Karl-Heinz Schleifer, and Erko Stackebrandt, 90-98. Springer, New York (Borenshtein and Schauer, 2006, 91-93).

[28] Liebl, Wolfgang. (2006). "*Corynebacterium* - Nonmedical". In *The Prokaryotes - A Handbook on the Biology of Bacteria*, vol. *3* (3rd edition), edited by Martin Dworkin, Stanley Falkow, Eugene Rosenberg, Karl-Heinz Schleifer, and Erko Stackebrandt, 796-818, Springer, New York (Liebl 2006, 796-97).

[29] Cabral, João P. S. (2010). "Water Microbiology. Bacterial Pathogens and Water". *International Journal of Environmental Research and Public Health*, *7* (10), 3657–3703. doi: 10.3390/ijerph7103657 (Cabral 2010, 3679).

[30] Grimont, Francine. & Patrick, A. D. Grimont. (2006). "The Genus *Enterobacter*". In *The Prokaryotes - A Handbook on the Biology of Bacteria*, vol. 6 (3rd edition), edited by Martin Dworkin, Stanley Falkow, Eugene Rosenberg, Karl-Heinz Schleifer, and Erko Stackebrandt, 197-214, Springer, New York (Grimont and Grimont 2006, 206-210).

[31] Welch, Rodney A. (2006). "The Genus *Escherichia*". In *The Prokaryotes - A Handbook on the Biology of Bacteria*, vol. *6* (3rd edition), edited by Martin Dworkin, Stanley Falkow, Eugene Rosenberg, Karl-Heinz Schleifer, and Erko Stackebrandt, 60-71, Springer, New York. (Welch, 2006, 62-63).

[32] Bernardet, Jean-François. & John, P. Bowman. (2006). "The genus *Flavobacterium*". In *The Prokaryotes - A Handbook on the Biology of Bacteria*, vol. *7* (3rd edition), edited by Martin Dworkin, Stanley Falkow, Eugene Rosenberg, Karl-Heinz Schleifer, and Erko

Stackebrandt, 481-531, Springer, New York (Bernardet and Bowman, 2006, 484, 493).

[33] McBee, Megan E. & David, B. Schauer. (2006). "The Genus *Hafnia*". In *The Prokaryotes - A Handbook on the Biology of Bacteria*, vol. *6* (3rd edition), edited by Martin Dworkin, Stanley Falkow, Eugene Rosenberg, Karl-Heinz Schleifer, and Erko Stackebrandt, 215-218, Springer, New York (McBee and Schauer 2006, 215-16).

[34] Hammes, Walter P. & Christian, Hertel. "The Genera *Lactobacillus* and *Carnobacterium". In *The Prokaryotes - A Handbook on the Biology of Bacteria*, vol. *4* (3rd edition), edited by Martin Dworkin, Stanley Falkow, Eugene Rosenberg, Karl-Heinz Schleifer, and Erko Stackebrandt, 320-403, Springer New York (Hammes 2006, 320-25).

[35] Kocur, Miloslav., Kloos, Wesley. E. & Schleifer, Karl-Heinz. (2006). "The Genus *Micrococcus*". In *The Prokaryotes - A Handbook on the Biology of Bacteria*, vol. *3* (3rd edition), edited by Martin Dworkin, Stanley Falkow, Eugene Rosenberg, Karl-Heinz Schleifer, and Erko Stackebrandt, 961-971, Springer, New York (Kocur et al. 2006, 961, 966).

[36] Betts, G. (2006). "Other Spoilage Bacteria". In *Food Spoilage Microorganisms*, edited by Clive de W. Blackburn, 668–693. Cambridge: Woodhead Publishing (Betts 2006, 688).

[37] Oldfield, N. J. & Ala'Aldeen, D. A. A. (2012). "*Neisseria* and *Moraxella*". In *Medical Microbiology* (18th ed.), edited by David Greenwood, Richard Slack, John Peitherer, and Mike Barer, 256–264, Elsevier (Oldfield 2012, 258).

[38] Moore, Edward R., Brian, J. Tindall., Martins, Dos Santos., Vitor, A. P., Dietmar, H. Pieper., Juan-Luis, Ramos. & Norberto, J. Palleroni. (2006). "Nonmedical: *Pseudomonas*". In *The Prokaryotes - A Handbook on the Biology of Bacteria*, vol. *6* (3rd edition), edited by Martin Dworkin, Stanley Falkow, Eugene Rosenberg, Karl-Heinz Schleifer, and Erko Stackebrandt, 646-703, Springer, New York (Moore et al. 2006, 648-654).

[39] Manos, Jim. & Robert, Belas. (2006). "The Genera *Proteus, Providencia*, and *Morganella*". In *The Prokaryotes - A Handbook on the Biology of Bacteria*, vol. *6* (3rd edition), edited by Martin Dworkin, Stanley Falkow, Eugene Rosenberg, Karl-Heinz Schleifer, and Erko Stackebrandt, 245-269, Springer, New York (Manos and Belas, 2006, 245-48).

[40] Wang, Yong. & Xiaoling, Pan. (2013). "*Proteus*". In *Encyclopedia of Food Safety*, edited by Y. Motarjemi, 486-489, Academic press, London, UK (Wang and Pan 2013, 486).

[41] Bowman, John P. (2006). "The Genus *Psychrobacter*". In *The Prokaryotes - A Handbook on the Biology of Bacteria*, vol. *6* (3rd edition), edited by Martin Dworkin, Stanley Falkow, Eugene Rosenberg, Karl-Heinz Schleifer, and Erko Stackebrandt, 920-930. Springer New York (Bowman 2006, 920-921).

[42] Kim, Su Jin., Seung, Chul Shin., Soon, Gyu Hong., Yung, Mi Lee., In-Geol, Choi. & Hyun, Park. (2012). "Genome Sequence of a Novel Member of the Genus Psychrobacter Isolated from Antarctic Soil". *Journal of Bacteriology*, *194*(9), 2403. doi: 10.1128/JB.00234-12 (Kim et al. 2012, 2403).

[43] Grimont, Francine. & Patrick, A. D. Grimont. (2006). "The Genus *Serratia*". In *The Prokaryotes - A Handbook on the Biology of Bacteria*, vol. *6* (3rd edition), edited by Martin Dworkin, Stanley Falkow, Eugene Rosenberg, Karl-Heinz Schleifer, and Erko Stackebrandt, 219-44, Springer, New York (Grimont and Grimont 2006, 230-235).

[44] Nealson, Kenneth H. & James, Scott. (2006). "Ecophysiology of the Genus *Shewanella*". In *The Prokaryotes - A Handbook on the Biology of Bacteria*, vol. *6* (3rd edition), edited by Martin Dworkin, Stanley Falkow, Eugene Rosenberg, Karl-Heinz Schleifer, and Erko Stackebrandt, 1133-1151, Springer, New York (Nealson and Scott 2006, 1133-1134).

[45] Zagorec, Monique. & Marie-Christine, Champomier-Verge. (2017). "Meat Microbiology and Spoilage", In *Lawrie's Meat Science*, edited

by Fidel Toldra, 187-203, Woodhead Publishing, Cambridge (Zagorec and Champomier-Verge 2017, 193-195).

[46] Nielsen, Dennis S., Tomas, Jacobsen., Lene, Jespersen., Anette, Granly Koch. & Nils, Arneborg. (2008). "Occurrence and Growth of Yeasts in Processed Meat Products - Implications for Potential Spoilage". *Meat Science*, *80*, 919-926. doi: 10.1016/ j.meatsci. 2008.04.011 (Nielsen et al. 2008, 919).

[47] Dillon, Vivian M. (1998). "Yeasts and Moulds Associated with Meat and Meat Products". In *The Microbiology of Meat and Poultry*, edited by Andrew Davies, and Ron Board, 85–117, Blackie Academic & Professional, New York (Dillon 1998, 92-94).

[48] Viljoen, B. C., Geornaras, I., Lamprecht, A. & Von Holy, A. (1998). "Yeast Populations Associated with Processed Poultry". *Food microbiology*, *15*(1), 113-117. doi: 10.1006/fmic.1997.0137 (Viljoen 1998, 113).

[49] International Commission on Microbiological Specifications for Foods (ICMSF). (1998). "Meat and Meat Products". In *Microorganisms in Foods 6 - Microbial Ecology of Food Commodities*, 1–74, Blackie Academic and Professional, London (ICMSF 1998, 1).

[50] Jackson, T. C., Acuff, G. R., Vanderzant, C., Sharp, T. R. & Savell, J. W. (1992). "Identification and Evaluation of Volatile Compounds of Vacuum and Modified Atmosphere Packaged Beef Strip Loins". *Meat Science*, *31*, 175-190. doi: 10.1016/0309-1740(92)90037-5 (Jackson et al. 1992, 175).

[51] Garcia-Lopez, M. L., Prieto, M. & Otero, A. (1998). "The Physiological Attributes of Gram-negative Bacteria Associated with Spoilage of Meat and Meat Products". In *The Microbiology of Meat and Poultry*, edited by Andrew Davies, and Ron Board, 1-35, Blackie Academic&Professional, New York (Garcia-Lopez 1998, 1, 4-6, 8).

[52] Cerveny, John., Joseph, D. Meyer. & Paul, A. Hall. (2009). "Microbiological Spoilage of Meat and Poultry Products". In *Compendium of the Microbiological Spoilage of Foods and*

Beverages, edited by Michael P. Doyl, 69-86, Springer, New York. (Cerveny 2009, 71-72, 75-76).

[53] Borch, Elisabeth., Marie-Louise, Kant-Muemans. & Ylva, Blixt. (1996). "Bacterial Spoilage of Meat and Cured Meat Products". *International Journal of Food Microbiology*, *33*, 103–120. doi: 10.1016/0168-1605(96)01135-X (Borch et al. 1996, 103).

[54] Barakat, R. K., Griffiths, M. W. & Harris, L. J. (2000). "Isolation and Characterization of *Carnobacterium*, *Lactococcus*, and *Enterococcus* spp. from Cooked, Modified Atmosphere Packaged, Refrigerated, Poultry Meat". *International Journal of Food Microbiology*, *62*, 83-94. doi: 10.1016/S0168-1605(00)00381-0 (Barakat 2000, 83).

[55] Ercolini, Danilo., Federica, Russo., Antonella, Nasi., Pasquale, Ferranti. & Francesco, Villani. (2009). "Mesophilic and Psychrotrophic Bacteria from Meat and their Spoilage Potential *in vitro* and in Beef". *Applied and Environmental Microbiology*, *75*, 1990–2001. doi: 10.1128/AEM.02762-08 (Ercolini 2009, 1992-1993).

[56] Casaburi, Annalisa., Paola, Piombino., George-John, Nychas., Francesco, Villani. & Danilo, Ercolini. (2015). "Bacterial population and volatilome associated to meat spoilage." *Food Microbiology*, *45*, 83-102. doi:10.1016/j.fm.2014.02.002 (Casaburi et al. 2015, 84).

[57] Russell, S. M., Fletcher, D. L. & Cox, N. A. (1995). "Spoilage Bacteria of Fresh Broiler Chicken Carcasses". *Poultry Science*, *74*(12), 2041–2047. doi:10.3382/ps.0742041 (Russell et al. 1995, 2041).

[58] Dainty, R. H. (1996). "Chemical-Biochemical Detection of Spoilage". *International Journal of Food Microbiology*, *33*, 19-34. doi: 10.1016/ 0168-1605(96)01137-3 (Dainty 1996, 19).

[59] Liao, C. H. (2006). "*Pseudomonas* and Related Genera". In *Food Spoilage Microorganisms*, edited by Blackburn, Clive de W., 507–540, Woodhead Publishing, Cambridge (Liao 2006, 513).

[60] Erkmen, Osman. & Faruk, T. Bozoglu. (2016). "Spoilage of Meat and Meat Products". In *Food Microbiology: Principles into Practice*,

edited by Osman Erkmen, and Faruk T. Bozoglu., 279-295, John Wiley & Sons (Erkmen and Bozoglu 2016, 283-284).

[61] Mead, G. (2004). "Shelf-life and Spoilage of Poultry Meat". In *Poultry meat processing and quality*, edited by G. Mead, 283-303 Elsevier (Mead 2004, 284).

[62] Schwaiger, Karin., Sabine, Huther., Christina, Hölzel., Peter, Kämpf. & Johann, Bauer. (2012). "Prevalence of Antibiotic-Resistant Enterobacteriaceae Isolated from Chicken and Pork Meat Purchased at the Slaughterhouse and at Retail in Bavaria, Germany". *International journal of food microbiology*, *154*(3), 206-211. doi: 10.1016/j.ijfoodmicro.2011.12.014 (Schwaiger et al. 2012, 206).

[63] Ismail, S. A. S., Deak, T., El-Rahman, H. A., Yassien, M. A. M. & Beuchat, L. R. (2000). "Presence and Changes in Populations of Yeasts on Raw and Processed Poultry Products Stored at Refrigeration Temperature". *International Journal of Food Microbiology*, *62*(1), 113-121. doi: 10.1016/S0168-1605(00)00414-1 (Ismail et al. 2000, 113).

[64] Nychas, G. J. E., Drosinos, E. H. & Board, R. G. (1998). "Chemical Changes in Stored Meat". *In The Microbiology of Meat and Poultry*, edited by A. Davies and R. Board, 288-326. Blackie Academic and Professional, London, England (Nychas et al. 1998, 316).

[65] Björkroth, Johanna K., Peter, Vandamme., & Hannu, J. Korkeala. (1998). "Identification and Characterization of *Leuconostoc carnosum*, associated with Production and Spoilage of Vacuum-Packaged, Sliced, Cooked Ham". *Applied and Environmental Microbiology*, *64*, 3313-3319. https://www.ncbi.nlm.nih.gov/pmc/articles/PMC106726/ Accessed September 28, 2017 (Björkroth et al. 1998, 3313).

[66] Borch, Elisabeth. & Halina, Agerhem. (1992). "Chemical, Microbial and Sensory Changes During the Anaerobic Cold Storage of Beef Inoculated with a Homofermentative *Lactobacillus* sp. or a *Leuconostoc* sp.". *International Journal of Food Microbiology*, *15*, 99-108. *doi:* 10.1016/0168-1605(92)90139-T (Borch and Agerhem 1992, 99).

[67] Chenoll, E., Macian, M. C., Elizaquivel, P. & Aznar, R. (2007). "Lactic acid Bacteria Associated with Vacuum-Packed Cooked Meat Product Spoilage: Population Analysis by rDNA-Based Methods". *Journal of Applied Microbiology*, *102*(2), 498-508. doi: 10.1111/ j.1365-2672.2006.03081.x (Chenoll et al. 2007, 503).

[68] Björkroth, Johanna K., Rolf, Geisen., Ulrich, Schillinger., Norbert, Weiss., Paul, De Vos., Wilhelm, H. Holzapfel., Hannu, J. Korkeala. & Peter, Vandamme. (2000). "Characterization of *Leuconostoc gasicomitatum* sp. nov., Associated with Spoiled Raw Tomato-Marinated Broiler Meat Strips Packaged under Modified-Atmosphere Conditions". *Applied and Environmental Microbiology*, *66*(9), 3764–72. http://aem.asm.org/content/66/9/3764.long Accessed September 28, 2017 (Björkroth et al. 2000, 3764).

[69] Gribble, Amanda. & Gale, Brightwell. (2013). "Spoilage Characteristics of *Brochothrix thermosphacta* and *campestris* in Chilled Vacuum Packaged Lamb, and their Detection and Identification by Real Time PCR". *Meat Science*, *94*, 361-368. doi: 10.1016/j.meatsci.2013.03.016 (Gribble and Brightwell 2013, 361).

[70] Gribble, Amanda., John, Mills. & Gale, Brightwell. (2014). "The spoilage Characteristics of *Brochothrix thermosphacta* and Two Psychrotolerant Enterobacteriaceae in Vacuum Packed Lamb and the Comparison between High and Low pH Cuts". *Meat Science*, *97*, 83-92. doi: 10.1016/j.meatsci.2014.01.006 (Gribble et al. 2014, 83).

[71] Kabisch, Jan., Constanze, Honing., Christina, Bohnlein., Rohtraud, Pichner., Gareis, M. & Manfred, Wenning. (2013). "*Kazachstania psychrophila* sp. nov., a Novel Psychrophilic Yeast Isolated from Vacuum-Packed Beef". *Antonie Van Leeuwenhoek*, *104*, 925-931. doi: 10.1007/s10482-013-0011-5 (Kabisch et al. 2013, 925).

[72] Nagy, E., Dlauchy, D., Medeiros, A. O., Peter, G. & Rosa, C. A. (2014). "*Yarrowia porcina* sp. nov. and *Yarrowia bubula* f.a. sp. nov., Two Yeast Species from Meat and River Sediment". *Antonie Van Leeuwenhoek*, *105*, 697-707. doi: 10.1007/s10482-014-0125-4. (Nagy et al. 2014, 697).

[73] Bisen, Prakash S., Mousumi, Debnath. & Godavarthi, B. K. S. Prasad. (2012). "Microbes as a Tool for Industry and Research". In *Microbes: Concepts and Applications*, edited by Prakash S. Bisen, Mousumi Debnath, and Godavarthi B.K.S. Prasad, 567-684. John Wiley and Sons (Bisen et al., 652-653).

[74] Ababouch, Lahsen. (2014). "Spoilage Problems Associated with Canning" In *Encyclopedia of Food Microbiology*, edited by Carl A. Batt, 175-180. Academic press, London, UK (Ababouch 2014, 178-180).

[75] Arnau, J., Serra, X., Comaposada, J., Gou, P. & Garriga, M. (2007). "Technologies to Shorten the Drying Period of Dry-Cured Meat Products". *Meat Science*, *77*(1), 81-89. doi: 10.1016/ j.meatsci. 2007.03.015 (Arnau 2007, 82).

[76] Battilani, Paola., Amedeo, Pietri., Paola, Giorni., Silvia, Formenti., Terenzio, Bertuzzi., Tania, Toscani., Roberta, Virgili. & Zofia, Kozakiewicz. (2007). "*Penicillium* Populations in Dry-Cured Ham Manufacturing Plants". *Journal of Food Protection*, *70*(4), 975-980. https://www.ncbi.nlm.nih.gov/pubmed/17477269 (Battilani et al. 2007, 975).

[77] Sørensen, Louise Marie., Tomas, Jacobsen., Perr, Væggemose Nielsen., Jens, Christian Frisvad. & Anette, Granly Koch. (2008). "Mycobiota in the Processing Areas of Two Different Meat Products". *International Journal of Food Microbiology*, *124*(1), 58-64. doi: 10.1016/j.ijfoodmicro.2008.02.019 (Sørensen et al. 2008, 58).

[78] Sonjak, Silva., Mia, Ličen., Jens, Christian Frisvad. & Nina, Gunde-Cimerman. (2011). "The Mycobiota of Three Dry-Cured Meat Products from Slovenia". *Food microbiology*, *28*(3), 373-376. doi: 10.1016/j.fm.2010.09.007 (Sonjak et al. 2011, 373).

[79] Asefa, Dereje T., Ragnhild, O. Gjerde., Maan, S. Sidhu., Solveig, Langsrud., Cathrine, F. Kure., Truls, Nesbakken. & Ida, Skaar. (2009). "Moulds Contaminants on Norwegian Dry-cured Meat Products". *International Journal of Food Microbiology*, *128*(3), 435-439 doi: 10.1016/j.ijfoodmicro.2008.09.024 (Asefa 2009, 435).

[80] Lozano-Ojalvo, Daniel., Alicia, Rodríguez., Mirian, Cordero., Victoria, Bernáldez., Mariana, Reyes-Prieto. & Juan, J. Córdoba. (2015). Characterisation and Detection of Spoilage Mould Responsible for Black Spot in Dry-Cured Fermented Sausages". *Meat Science*, *100*, 283-290. doi: 10.1016/j.meatsci.2014.10.003 (Lozano-Ojalvo et al. 2015, 283).

[81] Martin, Alberto., María, J. Benito., Alejandro, Hernandez., Francisco, Pérez-Nevado., Juan, J. Córdoba. & María, G. Córdoba. (2008). "Characterisation of Microbial Deep Spoilage in Iberian Dry-Cured Ham". *Meat Science*, *78*(4), 475-484. doi: 10.1016/j.meatsci. 2007. 07.017 (Martin et al. 2008, 475).

[82] Andrade, Maria J., Elena, Rodas., Ana, Durbán., Andres, Moya. & Juan, J. Córdoba. (2012). "Characterization and Control of Microbial Black Spot Spoilage in Dry-Cured Iberian Ham". *Food Control*, *23*(1), 128-36. doi: 10.1016/j.foodcont.2011.06.023 (Andrade 2012, 135).

[83] Champomier-Vergès, Marie-Christine. & Monique, Zagorec. (2015). "Spoilage Microorganisms: Risks and Control". In *Handbook of Fermented Meat and Poultry*, edited by Fidel Toldra, 383-388. John Wiley & Sons, Ltd. (Champomier-Vergès and Zagorec 2015, 385).

[84] Dave, D. & Ghaly, A. E. (2011). "Meat Spoilage Mechanisms and Preservation Techniques: A Critical Review". *American Journal of Agricultural and Biological Sciences*, *6*(4), 486-510. doi: 10.3844/ ajabssp.2011.486.510 (Dave 2011, 487).

[85] Nychas, G. J. E. & Drosinos, E. H. (2014). "Spoilage of Meat". In *Encyclopedia of food microbiology*, edited by Carl A. Batt, 514-519, Academic press, London, UK (Nychas and Drosinos 2014, 516-517).

[86] Sofos, John N., George, Flick., George-John, Nychas., Corliss, A. O'Bryan., Steven, C. Ricke. & Philip, G. Crandall. (2015). "Meat, Poultry and Seafood". In *Food Microbiology Fundamentals and Frontiers*, edited by Michael P. Doyle, and Robert L. Beuchat, 111-167. ASM Press, Washington, DC (Sofos et al. 2015, 116-117).

[87] Jones, Rhys J. (2004). "Observations on the Succession Dynamics of Lactic Acid Bacteria Populations in Chill-Stored Vacuum-Packaged

Beef". *International Journal of Food Microbiology*, *90*(3), 273-282 doi:10.1016/S0168-1605(03)00310-6 (Jones 2004, 273).

[88] Andersen, H. J. (2014). "Bacon Production". In Devine, C., and Dikeman, M. (eds). *Encyclopedia of Meat Sciences*, 51-56. Elsevier (Andersen 2014. 55).

[89] Húngaro, Humberto M., Magdevis, Y. R., Caturla, Cláudia N. Horita., Marianna, M. Furtado. & Anderson, S. Sant'Ana. (2016). "Blown Pack Spoilage In Vacuum-Packaged Meat: A Review on Clostridia as Causative Agents, Sources, Detection Methods, Contributing Factors and Mitigation Strategies". *Trends in Food Science & Technology*, *52*, 123-138. doi: 10.1016/j.tifs.2016.04.010. (Húngaro et al. 2016, 123).

[90] Sakala, R. M., Hayashidani, H., Kato, Y., Kaneuchi, C. & Ogawa, M. (2002). Isolation and Characterization of *Lactococcus piscium* Strains from Vacuum-Packaged Refrigerated Beef". *Journal of Applied Microbiology*, *92*, 173-179. https://www.ncbi.nlm.nih.gov/ pubmed/ 11849342 Accessed September 14, 2017 (Sakala et al. 2002, 173).

[91] Lowry, P. D. & Gill, C. O. (1984). "Temperature and Water Activity Minima for Growth of Spoilage Moulds from Meat". *Journal of Applied Microbiology*, *56*(2), 193-199. doi: 10.1111/j.1365-2672. 1984.tb01339.x (Lowry and Gill 1984, 193).

[92] Gill, C. O., Lowry, P. D. & Margaret, E. Di Menna. (1981). "A Note on the Identities of Organisms Causing Black Spot Spoilage of Meat". *Journal of Applied Microbiology*, *51*(1), 183-187. doi: 10.1111/j.1365-2672.1981.tb00922.x (Gill et al. 1981, 183).

[93] Nørrung, Birgit., Kirk Andersen, J. & Sava, Buncic. (2009). "Main Concerns of Pathogenic Microorganisms in Meat". In *Safety of meat and processed meat*, edited by Fidel Toldrá, 3-29. Springer, New York (Nørrung et al. 2009, 3-12).

[94] Mor-Mur, Montserrat. & Josep, Yuste. (2010). Emerging Bacterial Pathogens in Meat and Poultry: An Overview. *Food and Bioprocess Technology*, *3*(1), 24-35. doi: 10.1007/s11947-009-0189-8 (Mor-Mur and Yuste 2010, 26).

[95] Montville, Thomas J. & Karl, R. Matthews. (2008). "*Salmonella* Species". In *Food microbiology: an introduction*, edited by Thomas J. Montville, and Karl R. Matthews, 97-111. ASM Press, Washington, D.C. (Montville and Matthews 2008, 101).

[96] Food and Drug Administration (FDA). (2011). "Bacterial Pathogen Growth and Inactivation. Appendix 4". In *Fish and Fishery Products Hazards and Controls Guidance*, 4th ed., edited by Leonard Barry, 417-438. Food and Drug Administration, Center for Food Safety and Applied Nutrition, Office of Seafood, Washington, DC. https:// www.fda.gov/downloads/food/guidanceregulation/ucm252447.pdf Accessed October 12, 2017 (FDA 2011, 420-421).

[97] Franklin, Kristyn., Erika, J. Lingohr., Catherine, Yoshida., Muna, Anjum., Levente, Bodrossy., Clifford, G. Clark., Andrew, M. Kropinski. & Mohamed, A. Karmali. (2011). "A Rapid Geno-Serotyping Tool for the Classification of *Salmonella* Serovars". *Journal of Clinical Microbiology*, *49*(8), 2954-65. doi: 10.1128/ JCM.02347-10 (Franklin 2011, 2954).

[98] Khaleque, A. & Bari, M. L. (2015). "Facts on Foodborne Pathogens". In *Foodborne Pathogens and Food Safety*, edited by M.L. Bari, and D. O. Ukuku, 35-81, CRC Press (Khaleque and Bari 2015, 39, 44, 46).

[99] Cox, J. M. & Pavic, A. (2014). "*Salmonella"*. In *Encyclopedia of Food Microbiology*, edited by Carl A. Batt, 322-331. Academic press, London, UK (Cox and Pavic. 2014, 324-326).

[100] Uzzau, S., Brown, D. J., Wallis, T., Rubino, S., Leori, G., Bernard, S., Casadesús, J., Platt, D. J. & Olsen, J. E. (2000). "Host Adapted Serotypes of *Salmonella enterica". Epidemiology and Infection*, *125*(2), 229-255, https://www.ncbi.nlm.nih.gov/pmc/articles/ PMC2 869595 Accessed October 6, 2017.

[101] Jay, J. M. (2005). "Foodborne Gastroenteritis Caused by *Salmonella* and *Shigella*". In: *Modern Food Microbiology*, 7th edition, edited by J. M. Jay, 619-636, *Springer Science and Business Media*, (Jay 2005, 620).

[102] World Health Organization (WHO). (2017). "*Salmonella* (non-typhoidal)". Fact sheet Reviewed, September 2017, http:// www. who.int/mediacentre/factsheets/fs139/en Accessed October 24, 2017.

[103] Blackburn, Clive de W. & Peter, McClure. (2009). "*Campylobacter* and *Arcobacter*". In *Foodborne Pathogens: Hazards, Risk Analysis and Control*, 2nd edition, edited by Clive de W. Blackburn, and Peter McClure, 718-762. Woodhead Publishing Ltd., Cambridge, UK (Blackburn and McClure 2009, 718).

[104] Rowe, M. T. & Madden, R. H. (2014). "*Campylobacter". In Encyclopedia of Food Microbiology*, edited by Carl A. Batt, 351-366, Academic press, London, UK (Rowe and Madden. 2014, 353).

[105] Lianou, Alexandra., Panagou, Efstathios Z. & Nychas, George-John E. (2017). "Meat Safety - I Foodborne Pathogens and Other Biological Issues". In *Lawrie's Meat Science*, edited by Toldra, Fidel, 521-52. Woodhead Publishing, Cambridge (Lianou 2017, 523-538).

[106] Lianou, A. & Sofos, J. N. (2007). "A Review of the Incidence and Transmission of *Listeria monocytogenes* in Ready-To-Eat Products in Retail and Food Service Environments". *Journal of Food Protection*, 70, 2172-2198. https://www.ncbi.nlm.nih.gov/ pubmed/ 17900099 Accessed October 21, 2017 (Lianou and Sofos 2007, 2172).

[107] Bell, C. & Kyriakides, A. (2009). "Pathogenic *Escherichia coli*". In *Foodborne Pathogens: Hazards, Risk Analysis and Control*, 2nd edition, edited by Clive de W. Blackburn, and Peter McClure, 581-626, Woodhead Publishing Ltd., Cambridge, UK (Bell and Kyriakides 2009, 581, 584).

[108] Beauchamp, Catherine S. & John, N. Sofos. (2010). "Diarrheagenic *Escherichia coli*". In *Pathogens and Toxins in Foods: Challenges and Interventions*, edited by Vijay K. Juneja, John N. Sofos, 71-94. ASM Press, Washington (Beauchamp and Sofos, 2010, 71-72, 76-77).

[109] Rhoades, J. R., Duffy, G. & Koutsoumanis, K. (2009). "Prevalence and Concentration of Verotoxigenic *Escherichia coli*, *Salmonella* and

Listeria monocytogenes in the Beef Production Chain: A Review". *Food Microbiology*, *26*, 357-376. doi: 10.1016/j.fm.2008.10.012 (Rhoades et al. 2009, 359).

[110] World Health Organization (WHO). (2017). *"Botulism". Fact sheet* Reviewed October 2017, http://www.who.int/mediacentre/ factsheets/fs270/en/ Accessed October 24, 2017.

[111] EFSA-ECDC (European Food Safety Authority-European Centre for Disease Prevention and Control). (2015). "The European Union Summary Report on Trends and Sources of Zoonoses, Zoonotic Agents, and Food-Borne Outbreaks in 2014". *EFSA Journal*, *13*(12), 4329, 191 pp. doi: 10.2903/j.efsa.2015.4329 (EFSA-ECDC 2015, 24).

[112] Gillaspy, A. F. & Iandolo, J. J. (2014). *"Staphylococcus".* In *Encyclopedia of Food Microbiology*, edited by Carl A. Batt, 482-507. Academic press, London, UK (Gillaspy and Iandolo 2014, 484-485).

[113] European Commission (EC). (2005). "Commission Regulation (EC) No 2073/2005 of 15 November 2005 on Microbiological Criteria for Foodstuffs". *Official Journal of the European Union*, L *338*, 1-26. http://data.europa.eu/eli/reg/2005/2073/oj Accessed November 14, 2017 (EC 2005, 9-10, 15-16).

In: Microbial Contamination of Food Product ISBN: 978-1-53613-524-4
Editors: G. Mocanu and O. Constantin

Chapter 2

MICROBIAL CONTAMINANTS OF FISH AND FISH PRODUCTS

Angelica Docan and Iulia Rodica Grecu*
Department of Food Science and Engineering,
Biotechnology and Aquaculture,
Faculty of Food Science and Engineering,
Dunărea de Jos University of Galați, Galați, România

ABSTRACT

The microbial quality of fish and fish production is a significant aspect of food safety. Fish is a highly perishable food that requires proper handling and preservation. Spoilage and pathogenic bacteria associated with fish, and fishery products can be classified into three general groups: indigenous bacteria belonging to the natural commensal microflora of fish, exogenous enteric bacteria from surrounding polluted water, and bacterial contamination during processing, storage or preparation for consumption. It is considered that microbial spoilage is a major cause in transforming the fish muscle from acceptable to unacceptable raw material for food industry. The pathogenic bacterial contamination of fish appears as a result of attacks on fish by undesirable microorganisms from external sources. Temperature and pH are limitation parameters for the

* Corresponding author: Email: iulia.grecu@ugal.ro.

survival and growth of bacteria in fish products. Good hygienic practice is a measure to avoid contamination and to ensure the safety of fish and fish products.

Keywords: commensal microflora, microbial spoilage, *Clostridium botulinum, Vibrio parahaemolyticus, Salmonella* spp.

INTRODUCTION

Fish and fish products are considered foods with a high nutritional value and play an important part in a healthy diet, due to their content rich in high-quality protein, omega-3 and omega-6 polyunsaturated fatty acids with higher efficiency in the human body, vitamins (mainly A, D, B6, B12), minerals such as potassium, magnesium, phosphorus, and their low-fat content as compared to other meats [1, 2] able to provide the necessary energy to the body in low-calorie conditions. However, under certain conditions there is a risk of human illness caused by fish consumption, a risk that depends on several factors: geographical area, season, degree of industrialization, pollution, people in contact with fish and related environment, traditional food habits and the immune system status of the exposed individual.

The detection of microbial pathogens in food is the solution to the prevention and investigation of problems bound to health and safety, since food safety is becoming a global health concern, and foodborne diseases take a major toll on health [3]. The microbial quality of fish and fish production is a significant aspect of food safety. The investigation of foodborne pathogens in food plays an important role in preventing foodborne pathogen transmission [4]. Crude fish products have been reported as vectors for foodborne sickness [5].

As raw material after harvest, fish itself is considered a highly perishable food that can denature faster than terrestrial animal meat, becoming unacceptable to ingestion due to the complex spoilage process having occurred post-mortem due to the activity of endogenous enzymes, chemical constituents and microbial growth.

Therefore, rigorous handling and preservation techniques applied during post-harvest fish processing are required in order to maintain the quality and safety of fish products. According to FAO statistics [6], out of the total of more than 146 million tones of world fish production in 2014, 87% was destined for direct human consumption. In addition, as the predilect processing method, statistics mentioned that 67 million tons of the fish for direct human consumption was manufactured in the form of live, fresh or chilled fish (46%), 17 million tones in dried, salted, smoked or other cured forms (about 12%), 19 million tones in prepared and preserved forms (13%) and 44 million tones (about 30%) in frozen form. Therefore, taking into account the fact that almost 26% of the total fish production was frozen [6], it can be considered that this technique of temperature reduction is still the main method capable of eliminating degradation risks and allowing fish to be distributed and marketed worldwide.

Spoilage Microorganisms

The flesh muscle of alive and healthy fish is considered sterile and, in normal conditions, no bacteria or other microorganisms should be present. During the fish life, a complex community of microorganisms is present on its skin, gills and in its gastrointestinal tract, all being involved in different interaction relationships with the host. All these microorganisms make up the fish microbiota, and its structure is directly associated with the quality of the surrounding aquatic environment (freshwater or marine waters, open sea or coastal regions, level of eutrophication, degree of anthropogenic pollution, etc.).

Therefore, on the epithelial surfaces there are epiphytic bacteria or fungi that either use fish as live support, nourishing themselves with nutrients from the water, or use the slime secretions (mucus) as food at the same time, forming the commensal microbiota [7-8]. They are predominantly aerobic, although there are also facultative anaerobic bacteria, which are Gram-negative and opportunistic, reflecting the poikilothermic nature of fish with a wide temperature range for growth [9]. In consequence, much experimental research on mucosal microbiota reported a wide spectrum of taxonomic genera taking into account the geographical location, the season and methods of sampling and the protocol used for isolation [1, 7, 10]. Predominantly, the Gram-negative bacteria belong to the *Pseudomonas, Shewanella, Acinetobacter, Flavobacterium, Moraxella, Cytophaga* and *Vibrio* genera for the marine fish, and to *Aeromonas, Pseudomonas, Achromobacter, Flavobacterium, Shewanella, Alcaligenes* and *Serratia* genera for the fresh fish. Furthermore, the Gram-positive bacteria are associated with the *Micrococcus, Streptococcus, Bacillus* genera [9, 11].

Fish originating from oligotrophic and colder waters exhibit lower autochthonous and psychrotrophic bacterial numbers than fish from eutrophic and/or warm waters with mesophilic bacteria. Due to collecting techniques and analyzing procedures (standard culture-dependent methods *vs* molecular based methods) the mesophilic organisms can rapidly overgrow the psychrotrophic ones [1, 12] or the aerobic ones the strictly anaerobic ones. Comparing the microbiological methods, it is estimated that both the number and diversity of microflora are rather undervalued in classical methods, as Huber et al. [13] found by means of molecular methods that less than 2% of the intestinal microbiota of the rainbow trout is cultivable on agar media, and 50-90% of the total microbiota is represented by aerobic species.

Furthermore, the anthropomorphic origin contamination of the fish microflora can occur in surrounding waters or during harvest, processing and storage. For example, Ampofo and Clerk [14], when testing the microbiota of tilapia skin, found out that its composition changed from *Bacillus* spp., *Citrobacter* spp., *Edwardsiella* spp., *Enterobacter, Escherichia* spp., *Pseudomonas* spp., *Salmonella* spp., and *Streptococcus* spp.(control group), in accordance with the type of fertilizer applied in the pond, into the following: (1) *Bacillus* spp., *Pasteurella* spp., *Salmonella* spp., *Shigella* spp., *Streptococcus* spp., and *Yersinia* spp. for poultry manure; (2) *Bacillus* spp., *Proteus* spp., *Pseudomonas* spp., and *Streptococcus* spp. for pig manure; (3) *Bacillus* spp., *Flavobacterium* spp.,

Salmonella spp., *Shigella* spp., *Staphylococcus* spp., and *Streptococcus* spp. for blood meal.

In the past decades, more research has been focused on gastrointestinal microbiota due to its symbiotic role in epithelial renewal, enzyme production involved in digestion, as well as the synergic regulation of the immune system associated to the gastrointestinal tract [10, 15]. Apart from the microflora mentioned above, the digestive tract of marine fish contains a much higher proportion of *Vibrio* strains, among other bacteria formerly identified as A*chromobacter* spp., *Pseudomonas* spp., *Xanthomonas* spp., as well Gram-positive, anaerobic microflora represented by *Clostridium* spp. (*Cl. sporogenes, Cl. perfringens, Cl. bifermentans, Cl. tetani, Cl. tertium, Cl. lentoputrescens, Cl. botulinum*) and other spore-forming bacteria [9].

It has been suggested that some fish gut bacteria like *Lactobacillus* spp. and *Carnobacterium* spp. might have positive effects by enhancing the immunological reaction of the fish organism, ensuring competition for adhesion receptors in gut colonization of allochthonous pathogens, or their antagonistic exclusion through the activity of specific compounds as bacteriocins, H_2O_2, and organic acids (i.e., lactic acid) [16]. Species belonging to the genus *Lactobacillus* found in the carp intestine have been discovered be age-dependent, most frequently distributed throughout the intestine's whole length and in the wall of its mid part[17].The occurrence of lactobacilli groups in the normal intestinal flora was evinced in both freshwater and marine fish by several reviews [7, 10, 15, 18, 19].

Based on previous research, Clements et al. [20] point out that fish hindgut microorganisms are valid gut symbionts that may bring a significant contribution to the host metabolism, being responsible for protein uptake along the gut in fresh water omnivorous tilapia, and more quantitatively important in hindgut-fermenting marine herbivorous fishes.

In general, fish is free of pathogenic bacteria, but its intestine can be contaminated with allochthonous Enterobacteriaceae species, such as *Salmonella* spp. and *Shigella* spp., transmitted by the warm-blooded animals which are not considered part of the normal intestinal microflora of the hosts. The presence of some bacterial species (*Vibrio parahaemolyticus, Vibrio cholera, Escherichia coli, Staphylococcus aureus, Streptococcus iniae,* etc.) in the intestinal tract of fish point out that they originate from waters with effluents polluted by domestic discharges [21 - 22]. It has been found that the intestinal microflora of fish is strongly influenced by the host's feeding regime [23]. In cases of prolonged starvation of fish, the bacterial load of the intestine is strongly reduced from 10^7CFU/g in actively feeding species to very few in non-feeding fish [9].

Recent research has found that cellulose-degrading bacteria *Clostridium* spp., *Citrobacter* spp., and *Leptotrichia* spp. were dominant in the herbivores, while *Cetobacterium* spp. and protease-producing bacteria were dominant in the carnivores [24].

Sources and Type of Spoilage Microorganisms

After fish harvesting, physical, chemical and microbiological post-mortem changes begin to occur in the tissues, making the transition from a fresh food product to complete product degradation. The gradual transformation of the fish, as raw material, from acceptable to unacceptable, is produced by the action of the indigenous fish enzymes and microflora present on the skin, gills and intestine during the spoilage process. Huss [25] considers that spoilage bacteria are the specific group in the microflora that produce the off-odors and off-flavors associated with spoilage. In this regard, microbiological spoilage is derived from the action of the microbiota of indigenous fish, associated with the allochthonous ones from the surrounding environment, during harvesting, handling, packing and storage.

Therefore, in aerobic storage conditions, the flora predominantly consists of the same Gram-negative non-fermenting bacteria, but in different proportion with a total count increased from 10^5-10^6CFU/cm^2, found in fresh fish from commercial bottom catches, to 10^8-10^9CFU/cm^2 of skin or CFU/g of flesh, when the spoilage process is initiated [9]. The increasing number of bacteria is a direct consequence of the slime oversecretion on the skin surface after the fish death, associated with the autolysis process installed immediately after *rigor mortis,* in which enzyme activity breaks down the mucus into basic nutrients like mucopolysaccharide compounds, free amino acids, trimethylamine oxide and other components. As a matter of fact, during autolysis, great changes occur in tissues and the decomposition comprises proteins, lipids and carbohydrates, with different compound concentration [26]. The degradation of proteins creates low-molecular nitrogenous compounds that provide the favorable conditions for the growth of spoilage bacteria.

The bacteria resident on the skin or in the digestive tract could penetrate into the exposed muscle through puncture wounds, skin cuts and blood. Usually, during adequate handling and under chill storage conditions, microbial spoilage is limited to a surface layer, but in higher temperature conditions bacteria proliferate freely, penetrate the flesh surrounding the belly and significantly invade the tissues.

During microbial spoilage, there is a breakdown of valuable components (proteins, lipids) leading to the formation of new compounds, that are responsible for the alterations in odor, flavor and texture of the fish meat. These compositional changes are the major concerns of the freshness of marketable products.

Fish is spoiled at very different rates in correlation with the different textures of the outer integument. For example, species like *Merlangius merlangus* and *Gadus morhua* which have a very fragile integument spoil rapidly as compared to several flatfish that have a thick epidermis, with a dense slime layer that contains several antibacterial components like antibodies, complement and bacteriolytic enzymes [25].

It is obvious that the prevalence of some peculiar groups of spoilage bacteria will be favored by environmental conditions: storage temperature (unpreserved, chilled or freezing), decomposition level induced by enzymatic activity, flesh moisture content and pH, processing techniques (salting, brining, drying and smoking) for obtaining fish products.

Therefore, for unpreserved fish, bacterial spoilage is determined by Gram-negative bacteria, fermentative bacteria (such as Vibrionaceae), whereas psychrotolerant Gram-negative bacteria (such as *Pseudomonas* spp. and *Shewanella* spp.) are indicated for chilled fish spoilage [27]. Because fish flesh is low in carbohydrates (less than 0.5%) and only a small amount of lactic acid is produced after death, the post mortem pH remains high, which allows pH sensitive microorganisms, such as *Shewanella putrefaciens,* to grow in fish meat.

In packed cod, Dalgaard [28] identified the main spoilage agent to be *Photobacterium phosphoreum,* which in quantitative measurements of microbial spoilage activity (i.e., yield factor for production of trimethylamine/CFU) was able to determine detectable off-odor at 10^7CFU/g, as compared to *Shewanella putrefaciens* isolated from spoiled packed cod, which needs to reach the limit of 10^8-10^9CFU/g in order to produce off-odours. Analyzing the activity within the fish processing plants, Reynisson et al. [29] found a high bacterial diversity, with *Photobacterium phosphoreum* as the most apparent species, followed by genera such as *Flavobacterium* spp., *Psychrobacter* spp., *Chryseobacter* spp., *Acinetobacter* spp. and *Pseudoalteromonas* spp. Species from the *Photobacterium* spp., *Pseudoalteromonas* spp., *Pseudomonas* spp., *Shewanella* spp. genera were identified as active spoiling agents during fish storage, while other bacteria were associated with a lesser spoilage potential, such as *Flavobacterium* spp., *Psychrobacter* spp., *Chryseobacter* spp., *Acinetobacter* spp.

In a vast study regarding the microflora adhering to the processing equipment during production, including after the cleaning and disinfecting procedures, in four different processing plants with highly processed products, 1009 microorganisms were isolated by classical methods (from agar plates) and identified [30]. The microflora of cold-smoked salmon plants was dominated by *Pseudomonas* spp., genera of the Neisseriaceae, and Enterobacteriaceae families, *Corynebacterium* spp., *Acinetobacter* spp. and lactic acid bacteria, whereas the microflora in a plant processing semi-preserved herring consisted of *Pseudomonas* spp., *Alcaligenes* spp. and Enterobacteriaceae. In caviar processing species of *Psychrobacter, Staphylococcus* and yeasts were detected. It should be noted that after cleaning and disinfection, the dominant adhering microorganisms were pseudomonads and yeasts, independently of the processing microflora – an important aspect hinting to potentially contaminating bacteria in the final food products.

Biochemical Activity and Metabolites in Bacterial Spoilage

Fish spoilage is a complex process resulting from three basic mechanisms: enzymatic autolysis, oxidation and microbial growth [31].

Microbiological decomposition begins with enzymatic autolysis and penetration of spoilage bacteria into the tissues, a process that takes place simultaneously [25]. Even from the initial stages of spoilage, bacterial growth increases rapidly, producing many proteolytic and hydrolytic enzymes. It is considered that spoilage bacteria metabolism is a major cause that produces the final products responsible for the off-flavors, such as ammonia, hydrogen sulphide and mercaptans, indole, skatole, etc.

The internal breakdown of protein structure follows rigor mortis and the final products of protein hydrolysis (sulphur-containing amino acids, low-molecular nitrogenous components), under enzymatic action, are nourishment for the bacterial metabolism, leading to the formation of hydrogen sulphide, biogenic amines such as putrescine, histamine and cadaverine, and methylmercaptan. The volatile sulphur-compounds have unpleasant odors, and can be detected even at ppb levels, therefore even minimal quantities have a considerable effect on fish quality, indicating the unacceptable phase of spoiled fish. [25]. The components induced by microbial spoilage are summarized in Table 2.1.

Different authors [25, 31-32] pointed out that bacteria such as *Shewanella putrefaciens, Aeromonas* spp., psychrotolerant Enterobacteriaceae, *P. phosphoreum* and *Vibrio* spp. can obtain energy by reducing odorless trimethylamine oxide (TMAO) to smelling trimethylamine (TMA), thus creating the ammonia-like off-flavors. This process is most noticeable in conditions where oxygen is not present. In some species, a trimethylamine N-oxide aldolase (TMAOase) can break

Table 2.1. List of metabolites involved in bacterial activity (after Huss [25] and Ghaly et al. [31])

Storage temperature	Packaging atmosphere	Dominating microflora	Specific spoilage organisms	Spoilage compounds*
0°C	Aerobic	Gram-negative psychrotrophic, non-fermentative rods *(Pseudomonas* spp., *Shewanella. putrefaciens, Moraxella* spp., *Acinetobacter* spp.*)*	*Shewanella putrefaciens*	TMA, H_2S, CH_3SH, $(CH_3)_2S$, HX
			Pseudomonas spp.	Ketones, aldehydes, esters, non - H2S sulphides
0°C	Vacuum	Gram-negative rods; psychrotrophic or with psychrophilic character (*Shewanella putrefaciens, Photobacterium)*	*Shewanella putrefaciens*	TMA, H_2S, CH_3SH, $(CH_3)_2S$, HX
			Photobacterium phosphoreum	TMA, HX
0°C	Modified atmosphere packaging (CO_2 containing)	Gram-negative fermentative rods with psychrophilic character *(Photobacterium)* Gram-negative non-fermentative psychrotrophic rods (1-10% of flora; *Pseudomonas, Shewanella putrefaciens)* Gram-positive rods (Lactic acid bacteria)	*Photobacterium phosphoreum*	TMA, HX
5°C	Aerobic, Vacuum	Gram-negative psychrotrophic rods (Vibrionaceae, *Shewanella putrefaciens)*	*Aeromonas* spp.	NH_3, acetic acid, butyric acid and propionic acid
			Shewanella putrefaciens	TMA, H_2S, CH_3SH, $(CH_3)_2S$, HX
5°C	Modified atmosphere packaging (CO_2 containing)	Gram-negative psychrotrophic rods (Vibrionaceae)	*Aeromonas* spp.	NH_3, acetic acid, butyric acid and propionic acid
20-30°C	Aerobic	Gram-negative mesophilic fermentative rods (Vibrionaceae, Enterobacteriaceae)	Motile *Aeromonas* spp. *(A. hydrophila)*	NH_3, acetic acid, butyric acid and propionic acid

*TMA-Trimethylamine; H_2S-Hydrogen sulphide; CH_3SH-Methylmercaptan; $(CH_3)_2S$-Dimethylsulphide; HX-Hypoxanthine; NH_3-Ammonia down TMAO to dimethylamine (DMA) and formaldehyde. *Pseudomonas* spp. appeared to be responsible for sweet-rotten, fruity spoilage odors, while *Shewanella putrefaciens* led to H_2S production from L-cysteine, as well as methyilmercaptan and dimethyl sulphide, from methionine [9].

Another example of progressive enzymatic activity is the production of hypoxanthine in parallel with TMA development. Hypoxanthine could be formed both by the autolytic and bacterial degradation, and the rate of bacterial transformation of inosine or inosine mono-phosphate is higher than nucleotides autolysis.

Furthermore, the bacterial implication in the breakdown of the aminoacids present in fish muscle leads to the formation of primary amines, such as histamine production from histidine, arginine from glutamic acid, etc. This bacterial transformation may cause food poisoning in extreme cases. The activity of the spoilage bacteria could be responsible for the urea breakdown, with ammonia formation in some marine fish.

PATHOGENS IN FISH

It is known that fish is a highly perishable food which needs proper handling and preservation. Fish may be contaminated after capture or during transportation to retail markets. After contamination and microorganism replication, decay may take place and consumption becomes dangerous [33]. Since a superior quality of fish must be maintained for human consumption, the main concern in fish processing is to prevent fish deterioration and the microbial contamination of fish meat. Fish quality is influenced by many factors, such as source, cooling methods, processing, packaging and storage conditions. The quality and freshness of fish are rapidly deteriorated through microbial and biochemical mechanisms [34]. According to the International Commission on Microbiological Specifications for Foods, the limit for the total aerobic plate counts in fresh and frozen fish is 10^7CFU/g [35], whereas loads of 10^7-10^8CFU/g produce detectable organoleptic spoilage [36].

The microbial contamination of fish appears as a result of the attacks on fish by undesirable microorganisms (bacteria, mold, fungi, viruses or their toxins and by-products) from external sources. Pathogens could be transmitted to fish in the rearing tank or during processing under inadequate hygiene conditions [37]. Cooling and freezing are widely used methods for fish preservation.

Types of Pathogens

There is a wide range of microbiological pathogens, which can determine contamination in fish and fish processing industries. Pathogenic bacteria associated with fish and fishery product can be classified into three general groups [38]:

1. indigenous bacteria that belong to the natural fish microflora (*Clostridium botulinum,* pathogenic *Vibrio* spp., *Aeromonas hydrophila*);
2. nonindigenous bacteria (enteric bacteria) that are present due to fecal contamination (*Salmonella* spp., *Shigella* spp., the pathogenic *Escherichia coli, Staphylococcus aureus*);
3. bacterial contamination during processing, storage or preparation for consumption (*Bacillus cereus, Listeria monocytogenes, Staphylococcus aureus, Clostridium perfringens, Salmonella* spp.).

According to the European Food Safety Authority, pathogens such as *Salmonella* spp., *Yersinia* spp., *E. coli,* and *Listeria monocytogenes* are responsible for important foodborne outbreaks worldwide [39].

Clostridium botulinum is an anaerobic, Gram positive, rod-shaped, motile, spore-forming bacterium, with the ability to produce a potent neurotoxin, known as botulinum. This bacterium species only grows and produces toxins in the absence of oxygen [40]. *Cl. botulinum* is divided into four distinct phenotypic groups (I-IV); based on the serological antigenic specificity of the toxin produced by each strain; seven serotypes are recognized: A, B, C, D, E, F and G. Types A, B, E and F cause human botulism, type C and D cause in most cases botulism in animals. Type E has been involved in most of the botulism cases associated with the consumption of fish and fishery products [41]. *Cl. botulinum* type B causes botulism upon eating traditional fish salad [42]. All fresh fish products serve as a favorable environment for the growth and production of this toxin by non-proteolytic and psychrophilic strains (type B, E). The genes for types A, B, E, and F can now be readily differentiated using quantitative PCR [43].

Vibrio parahaemolyticus is a Gram-negative, rod-shaped, motile, facultative aerobic, non-spore forming bacterium, with a single, polar flagellum, found in saltwater and marine fish, which, when ingested, causes gastrointestinal illness in humans [44]. *V. parahaemolyticus* can be classified according to serotype, based on the presence of particular genes: *V. parahaemolyticus* strains are considered pathogenic if the thermostable direct haemolysin (THD) and/or TDH-related haemolysin (TRH) genes are present [45]. *V. parahaemolyticus* is a bacterium commonly found in estuaries and along marine coasts; its multiplication is favored by both the higher temperature and the presence of organic substances in the water, such as those from fish processing waste, residual or domestic effluents [46]. *V. parahaemolyticus* has been involved in human vibrioses associated with the consumption of insufficiently heat-treated fish or fish products [47].

Salmonella spp. belong to a genus of rod-shaped Gram-negative bacteria, that are non-spore-forming, predominantly motile, facultative anaerobic enterobacteria, capable of generating ATP with oxygen ("aerobically") when it is available. *Salmonella* spp. are intracellular pathogens and some serotypes cause diseases. They are mesophilic microorganisms and are mainly distributed in the gastrointestinal tracts of mammals, reptiles, birds and environments polluted with human or animal feces [48]. Although *Salmonella* was divided into over 2.500 serovars, only some of them have been identified as dominant in fish or the aquatic environment [49]. Fishery products have been recognized as a major carrier of food-borne pathogens, because the aquatic environments are important reservoirs of *Salmonella* [50].

Sources of Pathogens

The presence of different bacterial species, including human pathogenic bacteria in fish can be bound to the direct contact with contaminated water and ingestion of bacteria from sediments or contaminated feed. Certainly, the bacteria identified in fish reflect the conditions and safety of aquatic environments. The infection source may be fish kept as a food source and a hobby. Outbreaks usually occur due to the consumption of insufficiently heat-treated fish or fish products contaminated after/during their processing. Frozen fish and related products from seawater, intensive handling, long-time transport or cooking in fishing containers straight on the deck, contribute to their contamination with microorganisms [21].

People most often get infected: through contact with infected fish while handling them, with water or other constituents of the fish life environment; through processing fish in the food industry and preparation of dishes; orally by consumption of contaminated fish or related products, water or other constituents of water environment [51].

The main habitat of *Cl. Botulinum* is the soil, but it is also found in sewage, rivers, lakes, sea water, fresh meat, and fish [52]. A considerably higher prevalence of type E is seen in bottom-feeding fish species as compared to plankton-feeders, because the sea was the primary source of contamination [53]. The prevalence of *Clostridium botulinum* type E in raw fish varied from 10 to 40%, depending on the fish species studied [54]. The bacterium becomes a risk when processing practices are insufficient to eliminate botulinal spores from raw fish, particularly improper thermal processing [55]. Most critical are the hygienic condition for handling the product after smoking. Semi-preserved fish (salted, smoked, and marinated) can be dangerous for human consumption if it contains toxins previously formed in raw fish. The heat resistance of the spores is increased by a high fat and protein content, brining and dry environment.

Among the potentially pathogenic vibrios occurring naturally on fish and shellfish, *V. parahaemolyticus* is the most widespread. Vibrios are abundant in aquatic environments and these bacteria were also observed on the skin, gills and intestinal tracts of fish or shellfish. In comparison with water and sediment samples, fish intestines contained the highest number of *V. vulnificus* and *V. parahaemolyticus* [56]. Vibrios are associated with raw seafood as they belong to the indigenous microflora of the environment at the time of seafood capture or harvest. Healthy live fish is protected by its immune system, and consequently bacteria cannot develop in its meat. After the fish death, the immune system no longer functions and the bacteria are able to proliferate freely [27]. *Vibrio* spp. can contaminate fish and fish products during improper handling, long-time transport, evisceration, and there is also cross-contamination from raw materials. The prevalence of *Vibrio* spp. in fish and aquatic environments is influenced by water salinity and temperature; they were more frequently found in warm and water with a lower salinity [57]. Furthermore, an improper temperature control of storage favors bacterial proliferation.

Salmonella spp. have been isolated from fish and fishery product, although they are not psychrotrophic or indigenous to the aquatic environment [58]. Fresh fish or seafood like oysters or shrimp can contain *Salmonella* sp., particularly if they are captured from areas contaminated with fecal pollution (prior to harvest and during harvest) or processed, packed, stored, distributed under unsanitary conditions, and consumed raw or slightly cooked [59-60].

There are some pathways of contamination of aquaculture systems with *Salmonella* spp.: water run-off, animals (domestic animals, frogs, rodents, birds and reptiles), fertilization of ponds, contaminated feed and water sources, primary farm processing. Fish products can become contaminated with *Salmonella* spp. through the use of unsanitary ice, water, containers, and unhygienic handling practices [61]. *Salmonella* spp. can contaminate water sources because of poor sanitation and incorrect disposal of human and animal waste. These pathogenic bacteria can survive for 10 to 15 days in septic tank systems [62], but they have higher survival rates in aquatic environments: up to 54 days in water, and up to 119 days in sediment samples [63]. Various authors have shown that high lipid content in some fish species may favor the growth of *Salmonella*. Water temperature plays an important role in the long-term survival of *Salmonella* in the environment. The localization of *Salmonella* spp. in fish it is not uniform; most frequently, pathogens were isolated from the feeding tract, as compared to other parts of the fish, i.e., skin and gills [64].

Survival of Pathogens in Fish

Temperature and pH are limiting parameters for the survival of bacteria in fish products; the combination of heat and low pH used for processing and pasteurization of the fish viscera, incorporated into fish diets, should guarantee that the resulting fish food is safe against spreading infectious diseases [65]. From the microbiological point of view, fish and fish products are a high risk foodstuff group; particularly *Vibrio parahaemolyticus* and *Clostridium botulinum* type E are pathogenic bacteria associated with fish meat [21]. The control of foodborne botulism caused by *Cl. botulinum* is based almost entirely on the thermal destruction (heating) of the spores or the inhibition of spore germination, cell growth and toxins produced in foods. Although the spores of *Cl. botulinum* are relatively resistant to heat, the toxins are known to be relatively sensitive. Normal cooking procedures usually annihilate toxins in foods. Heating food to an internal temperature of 78°C for 1 minute has been proposed as a general rule for destroying toxins in foods [40]. At present, the inhibition of botulinum toxigenesis in fishery products relies almost exclusively on NaCl addition and refrigeration below 3°C [54]. The bacterium can grow at temperatures lower than 3.3°C, but its toxin is destroyed by heating at 70°C; therefore, any product that is properly cooked and eaten on the same day is considered safe. The main risks are the products that either can be eaten raw (e.g., smoked salmon), or need not be cooked by the consumer, (e.g., hot-smoked fish). In practice, therefore, if fish is adequately iced immediately after catching and kept below 4°C at all stages of processing until eaten, it will remain safe. The alteration of the chemical and microbiological composition of fish by the addition of enough salt or acetic acid can make the fish completely safe, although the high concentration required would not be acceptable to most consumers. However, a pH level of 4.5 in marinated products or a salt concentration of 3.0% in smoked fish products is usually acceptable and provides protection and extension of shelf life.

V. parahaemolyticus is a slightly halophilic bacterium: it grows optimally at NaCl concentrations ranging between 2 and 4%, and poor growth is exhibited in products with a NaCl concentration below 0.5%. The bacterium is inactivated quickly in distilled water and its growth is halted at levels of 10% NaCl. *V. parahaemolyticus* grows within a temperature range between 5 and 43°C, with an optimum value of 37°C. The optimum pH range for its growth is 7.8 to 8.6, though it may grow in the pH range of 4.8 to 11 [66]. All vibrios are sensitive to heat, therefore, cooling and refrigeration are critical control measures to prevent the growth of these pathogenic microorganisms.

Even if the open marine waters are free from *Salmonella* spp., this pathogenic microorganism can multiply and survive in the estuarine and tropical freshwater environments for some weeks [67]. *Salmonella* prefers to develop at 37°C, being relatively resistant to various environmental factors as compared to other Gram-negative bacteria. The limiting conditions for *Salmonella* growth are wide: temperature between 5°C and 47°C, maxim salt content of 4.0% to 5.0%, pH between a minimum of 3.7 to a maximum of 9.5, while the pH for the optimal growth is close to neutral, respectively 6.5-7.5 [68].

The good sanitary practices in aquaculture production and biosecurity measures must be able to minimize but not eliminate *Salmonella* from fish and fish products. According to the Food and Agriculture Organization of the United Nations [61] it is recommended to implement some important control measures to minimize the risk of contaminating aquaculture products with *Salmonella* spp.:

- the prevention of cross contamination,
- the design and construction of cages, nets and containers must allow for proper cleaning and disinfection,
- farm source water should be free from sewage contamination,
- ice should be manufactured from drinkable water and produced in sanitary conditions,
- harvesting equipment and utensils should be kept in clean condition,
- fish products should be cooled quickly and maintained at temperatures approximating that of melting ice,
- sorting, weighing, washing, and drainage should be carried out quickly and hygienically; fish products should be transported in containers with ice to ensure a temperature around 0°C during the whole period of transport,
- regular monitoring of staff to identify potential bearers of *Salmonella* spp.

Table 2.2. contains information on the maximum, cumulative time and internal temperature combinations for the exposure of fish and fishery products that, under ordinary circumstances, assure safer conditions against bacterial pathogens that are of the greatest concern in seafood processing.

Fish is frequently contaminated with foodborne pathogens such as *Salmonella, Vibrio* spp., *Clostridium botulinum* or *Yersinia* spp. reflecting the microflora of the aquatic environment. The contamination of the natural environment of fish can affect human health because fish and fish products may be a potential source of pathogenic bacteria. Different factors such as contaminated water sources, human activity or inadequate hygiene during fish capture, handling, and transportation could affect the prevalence of bacteria in fish and the aquatic environment. The risk of these pathogenic microorganisms is increased due to the specific abilities of these bacteria to survive in the aquatic environment. Good hygienic practices are a measure to avoid contamination and provide the safety of fish and fish products [70].

Table 2.2. Time and temperature guidance for controlling pathogen growth and toxin formation in fish and fishery products [69]

Potentially hazardous condition	Product temperature	Maximum cumulative exposure time
Germination, growth, and toxin formation by *Clostridium botulinum* type E	3.3-5°C 6-10°C 11-21°C Above 21°C	7 days 2 days 1 hours 6 hours
Growth of *Salmonella* spp.	5.2-10°C 11-21°C Above 21°C	2 days 5 hours 2 hours
Growth of V*ibrio parahaemolyticus*	5-10°C 11-21°C 22-27°C Above 27°C	21 days 6 hours 2 hours 1 hour
Growth of *Shigella* spp.	6.1-10°C 11-21°C Above 21°C	2 days 5 hours 2 hours
Growth and toxin formation by *Staphylococcus aureus*	7-10°C 11-21°C Above 21°C	14 days 12 hours 3 hours
Growth of *Yersinia enterocolitica*	-1.3-10°C 11-21°C Above 21°C	1 day 6 hours 2.5 hours
Growth of pathogenic strains of *Escherichia coli*	6.6-10°C 11-21°C Above 21°C	2 days 5 hours 2 hours
Growth of *Listeria monocytogenes*	-0.4-5°C 6-10°C 11-21°C 22-30°C Above 30°C	7 days 1 day 7 hours 3 hours 1 hour

ACKNOWLEDGMENTS

This work was supported by a grant from the Romanian National Authority for Scientific Research and Innovation, CNCS/CCCDI – UEFISCDI, project number PN-III-P2-2.1-BG-2016-0417, within PNCDI III

References

[1] Fernandes, Rhea. (2009). *Microbiology Handbook: Fish and Seafood*. Leatherhead Food International Ltd. Surrey, UK. http://www.leatherheadfood.com. Accessed December 10, 2017 (Rhea 2009, 1).

[2] Onurlubas, E. (2013). "The factors affecting fish consumption of the consumers in KeSan Township in Edirne". *Bulgarian Journal Science*, *19* (6), 1346-1350 (Onurlubas 2013, 1348).

[3] Zhao, Cuiwei., Beilei, Ge., Juan De, Villena., Robert, Sudler1., Emily, Yeh., Shaohua, Zhao., David G., White., David, Wagner. & Jianghong, Meng. (2001). "Prevalence of *Campylobacter* spp., *Escherichia coli* and *Salmonella* in Retail Chicken, Turkey, Pork, and Beef from the Greater Washington, DC, area". *Applied and Environmental Microbiology*, 67, 5431–5436. doi: 10.1128/AEM.67.12.5431-5436.2001 (Zhao et al. 2001, 5431).

[4] WHO. (2008). *World Health Organization*, Foodborne Disease Outbreaks: Guidelines for Investigation and Control. Geneva. http://www.who.int/foodsafety/publications/foodborne_disease/outbreak_guidelines.pdf. Accessed December 10, 2017 (WHO 2008, 1).

[5] Uradznski, J., Beata, Wysok. & Gomolka-Pawlicka, M. (2007). "Biological and Chemical Hazards Occurring in Fish and Fishery Products". *Polish Journal of Veterinary Sciences*, *10*, 183 – 188. https://www.ncbi.nlm.nih.gov/pubmed/17937192. Accessed November 28, 2017 (Uradznski et al. 2007, 183).

[6] *FAO*. (2016). The State of World Fisheries and Aquaculture 2016. Contributing to food security and nutrition for all. Rome. http://www.fao.org/3/a-i5555e.pdf. Accessed December 10, 2017 (FAO 2016, 46).

[7] Romero, Jaime, Einar, Ringø & Daniel L., Merrifield. (2014). "The Gut Microbiota of Fish". In *Aquaculture Nutrition: Gut Health, Probiotics and Prebiotics*, First Edition, edited by Daniel Merrifield, and Einar Ringø. John Wiley & Sons, Ltd. (Romero et al. 2014, 32).

[8] Kelly, Cecelia & Irene Salinas. (2017). "Under Pressure: Interactions between Commensal Microbiota and the Teleost Immune System". *Front. Immunol.*, *8*, 559. doi: 10.3389/fimmu.2017.00559 (Kelly and Salinas 2017, 2).

[9] Lougovois, V. P. & Kyrana, V. R. (2005). "Freshness Quality and Spoilage of Chill-Stored Fish". In *Food Policy, Control and Research*, edited by Arthur P. Riley, 35-86. Nova Science Publishers, Inc. (Lougovois and Kyrana 2005, 8).

[10] Wang, An R., Chao Ran, Einar Ringø & Zhi G. Zhou. (2017). "Progress in fish gastrointestinal microbiota research". *Reviews in Aquaculture*, 0, 1-15, doi: 10.1111/raq.12191 (Wang et al. 2017, 1).

[11] Ionescu, Aurelia, Margareta Zara, Gabriela Gurau, Iuliana Aprodu, Aida Vasile, Elpida Păltânea. (2006). *Procesarea industrială peştelui,* 91-106. Editura Fundaţiei Universitare "Dunărea de Jos" Galaţi [Fish Processing (Industrial)] (Ionescu et al. 2010, 91-92).

[12] Mahdi, Ghanbari, Wolfgang Kneifel. & Konrad J. Domig. (2015). "A New View of the Fish Gut Microbiome: Advances from Next-Generation Sequencing". *Aquaculture*, *448*, 464–475. doi: 10.1016/j.aquaculture.2015.06.033 (Mahdi Ghanbari et al. 2105, 464).

[13] Huber, I., Spanggaard, B., Appel, F. K., Rossen, L., Nielsen, T. & Gram, L. (2004). "Phylogenetic Analysis and *in situ* Identification of the Intestinal Microbial Community of Rainbow Trout (Oncorhynchus mykiss, Walbaum)". *Journal of Applied Microbiology*, *96*(1), 117–132. doi: 10.1046/j.1365-2672.2003.02109.x (Huber et al. 2004, 120).

[14] Ampofo, Joseph Addo, & George C. Clerk. (2010). "Diversity of Bacteria Contaminants in Tissues of Fish Cultured in Organic Waste-Fertilized Ponds: Health Implications". *The Open Fish Science Journal*, *3*, 142-146. doi: 10.2174/1874401X01003010142 (Ampofo and Clerk 2010, 143).

[15] Izvekova, G. I., Izvekov, E. I. & Plotnikov, A. O. (2007). "Symbiotic Microflora in Fishes of Different Ecological Groups". *Biology*

Bulletin, *34* (6):610-618. doi: 10.1134/S106235900706012X (Izvekova et al. 2007, 610).

[16] Ringø, Einar, & François-Joël Gatesoupe. (1998). "Lactic Acid Bacteria in Fish: A Review". *Aquaculture*, *160*(3-4), 177–203. doi: 10.1016/S0044-8486(97)00299-8 (Ringø 1998, 179).

[17] Jankauskiene, R. (2002). "Defence Mechanisms in Fish: Frequency of the Genus *Lactobacillus* Bacteria in the Intestinal Tract Microflora of Carp". *Biologija*, *2*, 13-17. http://elibrary.lt/resursai/LMA/Biologija/B-13-1.pdf. Accessed December 10, 2017 (Jankauskiene 2002, 13).

[18] Gatesoupe, François-Joël. (2008). "Updating the Importance of Lactic Acid Bacteria in Fish Farming: Natural Occurrence and Probiotic Treatments". *Journal of Molecular Microbiology and Biotechnology*, *14*(1-3), 107–114, doi: 10.1159/000106089 (Gatesoupe 2008, 110).

[19] Bucioa, Adolfo, Ralf, Hartemink., Johan W., Schrama., Johan, Verreth. & Frans M. Rombouts. (2006). "Presence of Lactobacilli in the Intestinal Content of Freshwater Fish from a River and from a Farm with a Recirculation System". *Food Microbiology*, *23*, 476–482 (Bucioa et al. 2006, 479).

[20] Clements, Kendall D., Esther R. Angert, W. Linn Montgomery, & Howard Choat, J. (2014). "Intestinal Microbiota in Fishes: What's Known and What's Not". *Molecular Ecology*, *23*, 1891-1898. doi: 10.1111/mec.12699 (Clements 2014, 1892).

[21] Novotny, L., Dvorska, L., Lorencova, A., Beran, V. & Pavlik, I. (2004). Fish: A Potential Source of Bacterial Pathogens for Human Beings". *Veterinarni medicina*, *49*(9), 343–358. https://www.vri.cz/docs/vetmed/49-9-343.pdf Accessed October 2, 2017 (Novotny et al. 2004, 344).

[22] Pal, Mahendra, Asnake Ketema, Manyazewal Anberber, Selamawit Mulu, & Yashodhara Dutta. (2016). "Microbial Quality of Fish and Fish Products". *Beverage & Food World*, *43*(2), 6–49. https://www.researchgate.net/publication/295857928. Accessed December 12, 2017 (Pal et al. 2016, 48).

[23] Ringø, E., Zhou, Z., Vecino, J. L. G., Wadsworth, S., Romero, J., Krogdahl, Å., Olsen, R. E., Dimitroglou, A., Foey, A., Davies, S., Owen, M., Lauzon, H. L., Martinsen, L. L., De Schryver, P., Bossier, P., Sperstad, S. & Merrifield, D. L. (2016). "Effect of Dietary Components on the Gut Microbiota of Aquatic Animals. A Never-Ending Story?" *Aquaculture Nutrition*, *22*, 219-282. doi:10.1111/anu.12346 (Ringø et al. 2016, 219).

[24] Liu, Han, Xianwu Guo, Ravi Gooneratne, Ruifang Lai, Cong Zeng, Fanbin Zhan & Weimin Wang. (2016). "The Gut Microbiome and Degradation Enzyme Activity of Wild Freshwater Fishes Influenced by their Trophic Levels". *Scientific Reports*, *6*, 1-12. doi: 10.1038/srep24340 (Liu et al. 2016, 8-9).

[25] Huss, Hans Henrik. (1995). "Postmortem Changes in Fish". In *Quality and Quality Changes in Fresh Fish,* edited by FAO Fisheries Technical Paper - 348, http://www.fao.org/docrep/v7180e/V7180E06.htm. Accessed December 6, 2017 (Huss 1995).

[26] Bykowski, P. & Dutkiewicz, D. (1996). "Freshwater Fish as Raw Material for Processing". In: *Freshwater Fish Processing and Equipment in Small Plants*. FAO Fisheries Circular No. 905. Rome. http://www.fao.org/docrep/w0495e/w0495e00.htm.). Accessed December 6, 2017 (Bykowski and Dutkiewicz 1996).

[27] Gram, Lone, & Huss Hans Henrik. (2000). "Fresh and Processed Fish and Shellfish". In: *The Microbiological Safety and Quality of Food*, edited by Barbara Lund, Anthony C. Baird-Parker, Grahame W. Gould, 472-506. Maryland: Aspen Publishers (Gram et al. 2000, 487).

[28] Dalgaard, Paw. (1995). "Qualitative and Quantitative Characterization of Spoilage Bacteria from Packed Fish". *International Journal of Food Microbiology*, *26*(3), 319-333. doi: 10.1016/0168-1605(94)00137-U (Dalgaard 1995, 319).

[29] Reynisson, Eyjólfur, Sveinn Haukur Magnússon, Árni Rafn Rúnarsson & Viggó Þór Marteinsson. (2010). "Bacterial Diversity in the Processing Environment of Fish Products". In *Matis - Food Research, Innovation & Safety.* Matís Report.

http://www.matis.is/media/matis/utgafa/11-10-Lokaskyrsla-1790.pdf. Accessed December 01, 2017 (Reynisson et al. 2010, 2).

[30] Bagge-Ravna, Dorthe, Yin Ng, Mette Hjelm, Jesper N. Christiansen, Charlotte Johansen, & Lone Gram. (2003). "The Microbial Ecology of Processing Equipment in Different Fish Industries-Analysis of The Microflora During Processing and Following Cleaning and Disinfection". *International Journal of Food Microbiology*, *87*, 239-250. doi: 10.1016/S0168-1605(03)00067-9 (Bagge-Ravna et al. 2003, 239).

[31] Ghaly, A. E., Dave, D., Budge, S. & Brooks, M. S. (2010). "Fish Spoilage Mechanisms and Preservation Techniques: Review". *American Journal of Applied Sciences*, *7*(7), 859-877. doi: 10.3844/ajassp.2010.859.877 (Ghaly 2010, 862).

[32] Gram, Lone, & Paw Dalgaard. (2002). "Fish Spoilage Bacteria - Problems and Solutions". *Current Opinion in Biotechnology*, *13*, 262-266. doi:10.1016/S0958-1669(02)00309-9. (Gram and Dalgaard 2002, 263).

[33] Alparslan, Yunus, Hatice Hasanhocaoglu, Cansu Metin, & Tacnur Baygar. (2014). "Determination of Meat Quality of Sea Bass (Dicentrarchus labrax) Sold at Different Selling Areas". *Emirates Journal of Food and Agriculture*, *26*(3), 293-301. doi: 10.9755/ejfa.v26i3.16528 (Alparslan et al. 2014, 293).

[34] Al-Jasser, Mohammed Saud, & Fahad Mohammed Al-Jasass. (2014). "Study the Chemical, Physical Changes and Microbial Growth as Quality Measurement of Fish". *Annual Research and Review in Biology*, *4*(9), 1406-1420. doi: 10.9734/ARRB/2014/7131 (Al-Jasser and Al-Jasass 2014, 1407).

[35] ICMSF (International Commission on Microbiological Specifications for Foods) (1986). "Sampling Plans for Fish and Shellfish". In *Microorganisms in Foods 2. Sampling for Microbiological Analysis: Principles and Scientific Applications.* 2nd edition, edited by ICMSF, 181-196. Toronto: University of Toronto Press (ICMSF 1986, 190).

[36] Broekaert K., Heyndrickx, M., Herman, L., Devlieghere, F. & Vlaemynck, G. (2011). "Seafood Quality Analysis: Molecular

Identification of Dominant Microbiota after Ice Storage on Several General Growth Media". *Food Microbiology*, *28*(6), 1162-1169. doi: 10.1016/j.fm.2011.03.009 (Broekaert et al. 2011, 1162).

[37] Uddin, G. M., Larsen, M. H., Guardabassi, L. & Dalsgaard, A. (2013). "Bacterial Flora and Antimicrobial Resistance in Raw Frozen Cultured Seafood Imported to Denmark". *Journal of Food Protection*, *76*(3), 490-499. doi: 10.4315/0362-028X.JFP-12-402 (Uddin et al. 2013, 490).

[38] Lyhs, Ulrike. (2009). "Microbiological Methods". In *Fishery Products Quality, Safety and Authenticity*, edited by Hartmut Rehbein, and Jörg Oehlenschläger, 318-348. Blackwell Publishing Ltd. (Lyhs 2009, 318).

[39] EFSA (European Food Safety Authority) and ECDC (European Centre for Disease Prevention and Control). (2015). "The European Union Summary Report on Trends and Sources of Zoonoses, Zoonotic Agents, and Food-Borne Outbreaks in 2013". *EFSA Journal*, *13*(1), 1-165. doi: 10.2903/j.efsa.2015.3991 (EFSA 2015, 35).

[40] Aberoumand, Ali. (2010). "Occurrence of *Clostridium botulinum* in Fish and Fishery Products in Retail Trade. A Review Article". *World Journal of Fish and Marine Science*, *2*(3), 246-250. https://www.idosi.org/wjfms/wjfms2(3)10/14.pdf Accessed September 12, 2017 (Aberoumand 2010, 246).

[41] Hauschild, Andreas H. W. (1989). "*Clostridium botulinum*". In *Foodborne Bacterial Pathogens*, edited by Michael P. Doyle, 111-189. CRC Press, New York (Hauschild 1989, 1370).

[42] Weber, J. Todd, Richard G. Hibbs Jr., Ahmed Darwish, Ban Mishu, Andrew L. Corwin, Magda Rakha, Charles L. Hatheway, Said El Sharkawy, Sobhi Abd El-Rahim, Mohammed Fathi Sheba Al-Hamd, James E. Sarn, Paul A. Blake, & Robert V. Tauxe. (1993). "A Massive Outbreak of Type E Botulism Associated with Traditional Salted Fish in Cairo". *The Journal of Infectious Diseases*, *167*, 451–454. https://www.jstor.org/stable/30113098. Accessed September 12, 2017 (Weber 1993, 451).

[43] Satterfield, B. A., Stewart, A. F., Lew, C. S., Pickett, D. O., Cohen, M. N., Moore, E. A., Luedtke, P. F., O'Neill, K. L. & Robiso, R. A. (2010). "A Quadruplex Real-Time PCR Assay for Rapid Detection and Differentiation of the *Clostridium botulinum* Toxin Genes A, B, E and F". *Journal Medical Microbiology*, *59*, 55-64. doi: 10.1099/jmm.0.012567-0 (Satterfield et al. 2010, 56).

[44] Ryan, Kenneth J. (2004). *"Vibrio, Campylobacter, and Helicobacter"*. In *Sherris Medical Microbiology*, 4th ed., edited by Kenneth J. Ryan, and C. George Ray, 373-384. McGraw-Hill Education, Inc. (Ryan 2004, 374).

[45] Drake, Stephenie L., Angelo DePaola, & Lee-Ann Jaykus. (2007). "An Overview of *Vibrio vulnificus* and *Vibrio parahaemolyticus*". *Comprehensive Reviews in Food Science and Food Safety*, *6*, 120–14. doi: 10.1111/j.1541-4337.2007.00022.x (Drake et al. 2007, 374).

[46] Munteanu, G. & Bogatu, D. (2004). *Tratat de ihtiopatologie*, 704-717, Excelsior Art Timisoara [The Handbook of Fish Diseases] (Munteanu and Bogatu 2003, 705).

[47] Gauthier, David T. (2015). "Bacterial Zoonoses of Fishes: A Review and Appraisal of Evidence for Linkages Between Fish and Human Infections," *Veterinary Journal*, *203*, 27–35. doi: 10.1016/j.tvjl.2014.10.028 (Gauthier 2015, 31).

[48] Saeed, A. Mahdi & Ritha Naji. (2007). "Infection Agents". In *Epidemiologic Principles and Food Safety*, edited by Tamar Lasky, 18-40. Oxford University Press (Saeed and Naji 2007, 22).

[49] Agbaje, M., Begum, R. H., Oyekunle, M. A., Ojo, O. E., Adembi, O. T. (2011). "Evolution of *Salmonella* Nomenclature: A Critical Note". *Folia Microbiologica*, *56*, 497–503. doi10.1007/s12223-011-0075-4 (Agbaje et al. 2011, 497).

[50] Upadhyay, Bishnu Prasad, Fuangfa Utrarachkij, Jarinee Thongshoob, Yuvadee Mahakunkijcharoen, Niracha Wongchinda, Orasa Suthienkul, & Srisin Khusmith. (2010). "Detection of *Salmonella inva* Gene in Shrimp Enrichment Culture by Polymerase Chain Reaction". *Southeast Asian Journal of Tropical Medicine and Public Health*, *41*(2), 426-435. www.tm.mahidol.ac.th/seameo/2010-41-

2/21-4694.pdf Accessed November 11, 2017 (Upadhyay et al. 2010, 427).

[51] Acha, Pedro N. & Boris Szyfres. (2003). *Zoonoses and Communicable Diseases Common to Man and Animals: Bacterioses and Mycoses*. 3rd ed., 1-384. Scientific and Technical Publication No. 580, Pan American Health Organization, Regional Office of the WHO, Washington, USA. http://apps.who.int/iris/bitstream/10665/165519/2/9275119936.pdf Accessed November 21, 2017 (Acha and Szyfres 2003).

[52] Haagsma, J. (1991). "Pathogenic Anaerobic Bacteria and the Environment". *Revue scientifique et technique* (International Office of Epizootics), 10, 749–764. https://www.ncbi.nlm.nih.gov/pubmed/1782427 Accessed December 6, 2017. (Haagsma 1991, 749).

[53] Huss, H. H. & Pedersen, A. (1979). "*Clostridium botulinum* in fish". *Nord Vet-Medical*, *31*, 214-221. https://www.ncbi.nlm.nih.gov/pubmed/382097. Accessed December 6, 2017 (Huss and Pedersen 1979, 214).

[54] Hyytia, E., Hielm, S. & Korkeala, H. (1998). "Prevalence of *Clostridium botulinum* Type E in Finnish Fish and Fishery Products". *Epidemiology and Infection*, *120*, 245–250. http://www.jstor.org/stable/3864513. Accessed December 6, 2017 (Hyytia et al. 1998, 246).

[55] Chattopadhyay, P. & Adhikari, S. (2014). "Fish - Catching and Handling". In *Encyclopedia of Food Microbiology*, edited by Carl A. Batt, 923-931. Academic press, London, UK (Chattopadhyay and Adhikari 2014, 928).

[56] Givens, C. E., Bowers, J. C., DePaola, A., Hollibaugh, J. T. & Jones, J. L. (2014). "Occurrence and Distribution of *Vibrio vulnificus* and *Vibrio parahaemolyticus* - Potential Role for Fish, Oyster, Sediment and Water". *Letters in Applied Microbiology*, *58*, 503–510. doi: 10.1111/lam.12226 (Givens 2014, 507).

[57] Huehn, Stephan, Christin Eichhorn, Sara Urmersbach, Janina Breidenbach, Silke Bechlars, Nadja Bier, Thomas Alter, Edda

Bartelt, Christina Frank, Boris Oberheitmann, Florian Gunzer, Nicole Brennholt, Simone Böer, Bernd Appel, Ralf Dieckmann, & Eckhard Strauch. (2014). "Pathogenic Vibrios in Environmental, Seafood and Clinical Sources in Germany". *International Journal Medical Microbiology*, *304*, 843–850. doi: 10.1016/j.ijmm.2014.07.010 (Huehn et al. 2014, 847).

[58] Mol, S., Cosansu, S., Alakavuk, D. U. & Ozturan, S. (2010). "Survival of *Salmonella enteritidis* during Salting and Drying of Horse Mackerel (*Trachurus trachurus*) Fillets". *International Journal of Food Microbiology*, *139*, 36-40 (Mol et al. 2010, 38).

[59] Kumar, Rakesh, Surendran, P. K. & Thampuran Nirmala. (2009). "Detection and Characterization of Virulence Factors in Lactose Positive and Lactose Negative *Salmonella* Serovars Isolated from Seafood". *Food Control*, *20*(4), 376-380. doi: 10.1016/ j.foodcont.2008.06.00 (Kumar 2009, 376).

[60] Norhana, M. N. Wan., Susan E. Poolec, Hilton C. Deeth, & Gary A. Dyke. 2010. "Prevalence, Persistence and Control of *Salmonella* and *Listeria* in Shrimp and Shrimp Products". *Food Control*, *21*(4), 343-361. doi: 10.1016/j.foodcont.2009.06.020 (Norhana 2010, 358).

[61] FAO. (2010). Report of FAO Expert Workshop on the Application of Biosecurity Measures to Control Salmonella Contamination in Sustainable Aquaculture. *FAO Fisheries and Aquaculture Report No. 937*, Mangalore, India, 19–21 January 2010. www.fao.org/ docrep/013/i1547e/i1547e00.pdf. Accessed December 10, 2017 (FAO 2010).

[62] Parker, William F. & Brian J. Mee. (1982). "Survival of *Salmonella* adelaide and Fecal Coliforms in Coarse Sands of the Swan Coastal Plain, Western Australia". *Applied Environmental Microbiology*, *43*, 981-986 (Parker and Mee, 1982, 981).

[63] Moore, Barry C., Edward Martinez, John M. Gay, & Daniel H. Rice. (2003). "Survival of *Salmonella enterica* in Freshwater and Sediments, and Transmission by the Aquatic Midge *Chironomus tentans* (Chironomidae: diptera)". *Applied Environmental*

Microbiology, *69*, 4556-4560. doi:10.1128/AEM.69.8.4556-4560.2003 (Moore et al. 2003, 4556).

[64] Hatha, A. A. M. & P. Lakshmanaperumalsamy, P. (1997). "Prevalence of *Salmonella* in Fish and Crustaceans from Markets in Coimbatore, South India". *Food Microbiology*, *14*, 111-116. http://dyuthi.cusat.ac.in/purl/3931. Accessed December 10, 2017 (Hatha and Lakshmanaperumalsamy 1997, 111).

[65] Whipple, M. J. & Rohovec, J. S. (1994). "The Effect of Heat and Low pH on Selected Viral and Bacterial Fish Pathogens". *Aquaculture*, *123*, 179-189. doi: 10.1016/0044-8486(94)90056-6 (Whipple and Rohovec 1994, 180).

[66] Chai, T. J. & Pace, J. L. (2001). "*Vibrio parahaemolyticus*". In *Foodborne Disease Handbook - Bacterial Pathogens*, 2nd Edition, edited by Y. H. Hui, M. D. Pierson, and J. R. Gorham, 407-437. Marcel Dekker, Inc., New York (Chai and Pace 2001, 407).

[67] Huss, Hans Henrik, & Lone Gram. (2004). "Characterization of Hazards in Seafood Assurance of Seafood Quality". In *Assessment and Management of Seafood Safety and Quality*, FAO Fisheries Technical Paper 444. http://www.fao.org/docrep/006/y4743e/y4743e00.htm. Accessed December 10, 2017 (Huss and Gram 2004).

[68] Olgunoğlu, İlkan Ali. (2012). "*Salmonella* in Fish and Fishery Products". In *Salmonella - A Dangerous Foodborne Pathogen*, edited by Barakat S.M. Mahmoud, Available from: http://www.intechopen.com/books/salmonella-a-dangerous-foodborne-pathogen/salmonella-in-fish-andfishery-products, Accessed December 10, 2017 (Olgunoğlu 2012).

[69] U.S. Department of Health and Human Services/Food and Drug Administration/Center for Food Safety and Applied Nutrition. (2011). *Fish and Fishery Products Hazards and Controls Guidance*. Fourth Edition – April 2011. https://www.fda.gov/downloads/Food/GuidanceRegulation/UCM251970.pdf Accessed December 10, 2017.

[70] Novoslavskij, Aleksandr, Margarita Terentjeva, Inga Eizenberga, Olga Valciņa, Vadims Bartkevičs, & Aivars Bērziņš. (2016). "Major Foodborne Pathogens in Fish and Fish Products: A Review". *Annals of Microbiology*, *66*, 1-15. doi: 10.1007/s13213-015-1102-5 (Novoslavskij 2016, 12).

In: Microbial Contamination of Food Product ISBN: 978-1-53613-524-4
Editors: G. Mocanu and O. Constantin

Chapter 3

MICROBIAL CONTAMINANTS OF EGGS AND EGG PRODUCTS

Oana Emilia Constantin*
Department of Food Science and Engineering,
Biotechnology and Aquaculture,
Faculty of Food Science and Engineering,
Dunărea de Jos University of Galați,
Galați, România

ABSTRACT

Eggs and egg products are a category of nutritionally important foods and are representing a significant part in human diet all around the word. They are widely used to obtain different food products such as salad dressings, ice cream, confectionery products. Eggs and egg products can be contaminated with spoilage and pathogenic microorganisms during processing steps, and therefore it is very important to identify and control the main factors which lead to changes in nutritional quality and egg safety.

Keywords: eggs, egg products, *Salmonella* spp., spoilage

---

* E-mail: econstantin@ugal.ro.

Introduction

Eggs have specific nutritional properties and a high biological activity, as compared with other foods. Because of its function to form a new organism, the egg concentrates all the necessary nutrients in ideal proportions and in quantities that ensure the normal evolution of the new organism.

The egg is a complete food with a high nutritional value, used as food and for dietary purposes in the normal diet. Its nutritional quality results from the chemical composition, which varies according to species, breed, and season. The approximate chemical composition of the egg is variable, and the whole egg, without shell, contains: water 76.0%, proteins 12.8%, lipids 10.2%, carbohydrates 0.3%. and ash 0.7% [1].

The egg content is mostly sterile when the egg is expelled. However, eggs may be contaminated by both spoilage microorganism and pathogenic bacteria. Eggs may be contaminated in two ways: externally through the shell, and internally, during the egg formation. Therefore, eggs can represent a vector of bacteria responsible for food disease in humans, particularly *Salmonella* [2].

Due to their properties they are used in many sectors of the food industry, e.g., the bakery industry, dairy industry, in various forms: whole eggs or as white or yolk, fresh, pasteurized, frozen or dried eggs.

The control of the microbiological quality of the egg or egg products is very important, especially because they are used in some thermally unprocessed food or slightly baked foods.

Egg Structure

The contamination and development of microorganisms are influenced by the egg composition and structure that are also responsible for the egg protection against spoilage microorganisms. The structure of the egg is shown in Figure 3.1.

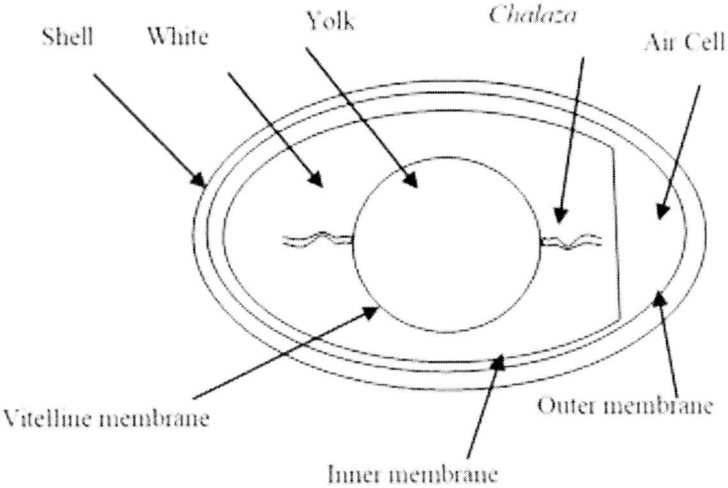

Figure 3.1. The egg structure.

The egg generally consists of shell and shell membranes, and two colloidal suspensions, with different composition:

- the egg white or albumen is formed of ovalbumin dispersed in a significant quantity of water;
- the egg yolk is a concentrated lipid emulsion which forms complex compounds with proteins.

Shell and Shell Membrane

The eggshell is a physical barrier able to prevent the contamination of the egg colloidal systems (albumen and yolk). The eggshell components are [3-5]:

- the cuticle, composed of glycoproteins, placed on the surface of the shell, which form a barrier able to slow down the water transfer from the egg to the environment, and vice versa. The cuticle is the first protection system against microorganisms.
- the calcareous layer, under the thin cuticle.

- the mammillary layer, consisting of compacted particles, which is placed between the calcareous layer and the outer membrane.

The shell membrane is made up of two layers of different sizes (48 and 22 μm, respectively), consisting of a protein-polysaccharide matrix [4].

The proteins found in the eggshell matrix are ovocalilin, ovocleidine, osteopontin, ovocalilamine, ovocalicin, clusterin, ovalbumin, lysozyme and ovotransferrin [6], and they have antimicrobial properties against both Gram negative and Gram-positive bacteria, probably because they disturb the bacterial membrane [7]. Cracks and fissures in the shell increase the penetration of microorganisms and may cause rapid egg damage. The contamination through the shell is also given by the number of pores on the shell, the age of the egg and the bird's condition [3].

Albumen

The egg white is an aqueous solution of different proteins, also containing very small amounts of lipids (0.03%), carbohydrates (about 1%), minerals (0.6%) and vitamins [4].

The pH of the fresh egg albumin is 7.6-7.9 and can increase with time and temperature to 9.7 during egg storage [8], and therefore limits the growth of bacteria in the egg white. Furthermore, microbiological spoilage is influenced by some albumen proteins with biological activity, e.g., enzyme (lysozyme); enzyme inhibitors (ovomucoid, ovoinhibitor) [4].

The egg white maintain the yolk in a central position away from the shell, and therefore the possibility of being contaminated is reduced.

Egg Yolk

The yolk is an oil-in-water emulsion with particles of different sizes, and is an excellent medium for the multiplication of microorganisms. The yolk is composed mainly of proteins and lipids, and small quantities of carbohydrates, minerals, vitamins, aroma substances, and colorants. This egg part has the pH = 6.9, may contain some substances (lipoproteins) with inhibitory effect [9], and is located at a great distance from the shell, and consequently from the microorganisms that can contaminate the egg trough shell cracks.

Microbial Contamination of Eggs

Egg contamination by microorganisms can be achieved in two ways: internal and external contamination [2, 6].

Internal contamination (vertical transmission) occurs during the egg formation in the oviduct and is usually related to *Salmonella* spp. Throughout this *transovarian route*, the egg white and yolk, are contaminated by the infected reproductive organs [10-11]. This contamination occurs when bacteria are found in high concentrations (10^8 - 10^9 cells per bird) [2].

External contamination (horizontal transmission) through the eggshell surface occurs after egg laying. Because after oviposition (egg expulsion), the egg has a higher temperature than the ambient air, and some pores of the shell are open [12], the bacteria from the environment can penetrate the shell. In a study conducted by De Reu et al. [13], the bacteria isolated throught the trans-shell route were: *Staphylococcus warneri*, *Serratia marcescens*, *Acinetobacter baumannii*, *Salmonella Enteritidis, Alcaligenes* spp., *Carnobacterium* spp., and *Pseudomonas* spp.

Spoilage Microorganisms

Shell Eggs

The microflora of eggs is mainly made up of Gram-positive bacteria, such as *Staphylococcus* spp., *Streptococcus* spp., *Aerococcus* spp., and *Micrococcus* spp. [14], which are located on the shell and come from the soil, and feces as primary sources of microorganisms [3], egg processing and handling equipment [15]. The Gram-negative bacteria are considered to be contaminants of the egg content, because they are considered to be more resistant to the egg's natural defense systems [14].

The bacteria responsible for egg spoilage are: *Pseudomonas* spp., *Alcaligenes* spp., *Aeromonas* spp., *Acinetobacter* spp., *Arthrobacter* spp., *Bacillus* spp., *Citrobacter* spp., *Cytophaga* spp., *Enterobacter* spp., *Escherichia* spp., *Flavobacterium* spp., *Hafnia* spp., *Micrococcus* spp., *Proteus* spp. and *Serratia* spp. [6]. Out of these, the main spoilers are *Pseudomonas* spp., followed by *Proteus* spp., *Alcaligenes* spp., *Aeromonas* spp., *Citrobacter* spp., and *Flavobacterium* spp. [8, 16-18].

The spoilage (rot) of shell eggs associated with the appearance of some coloring on the eggs are shown in Table 3.1.

Yeast and molds have a lower influence compared to bacteria in the contamination of eggs. In improper storage conditions, the eggshell may be contaminated by molds, especially in the air cell, that produce differently colored spots, such as: *Penicillium* spp. (blue, green and yellow spots), *Sporotrichum* spp. (pink spots), and *Cladosporium* spp. (black spots) [16]. Other molds responsible for egg spoilage are *Alternaria* spp., *Rhizopus* spp., *Mucor* spp., *Thamnidium* spp., *Botrytis* spp., and *Rhodotorula* spp. [15, 22].

Egg Products

Egg products are classified in figure 3.2 [4].

Liquid egg products are commercialized in various forms, such as yolk, egg white, whole eggs, mixed whole eggs, and are usually pasteurized, refrigerated or frozen until use. The initial microflora of liquid eggs is similar with the shell eggs, consisting of Gram negative and Gram-positive bacteria, but after pasteurization the bacteria number is reduced by 99% [22]. The species of *Alcaligenes* spp., *Proteus* spp., *Flavobacterium* spp. and *Bacillus* spp. can survive the pasteurization process.

Table 3.1. Bacteria associated with spoilage (rot) of shell eggs

Spoilage bacteria	Type of spoilage	Changes induced	Reference
Pseudomonas fluorescens	Green rot	Albumen greening Yolk may disintegrate	[16]
Proteus vulgaris	Black rot	Discoloration of yolk because of H_2S production Watery albumen which turns brown	[19-20]
Pseudomonas spp., *Aeromonas* spp., *Alcaligenes* spp., *Escherichia* spp., *Enterobacter* spp.	Black rot	Albumen digestion Yolk becomes black H_2S production or putrid odor	[14, 17]
Pseudomonas spp	Pink rot	Discoloration of albumen.	[19]
Proteus vulgaris *Proteus intermedium*	Custard rot	Yolk encrusted	[21]
Enterobacter spp.			[3]
Serratia marcescens	Red rot	Bright red albumen No smell	[19]
Acinetobacter spp., *Pseudomonas* spp., *Moraxella* spp., *Citrobacter* spp.	Colorless rot	Watery albumen Yolk disintegration. Possibly fruity odor	[14, 19]

Frozen egg products are usually obtained by pasteurizing the yolk, albumen or egg mixtures at 63°C for 1 min and then rapidly applying a low temperature treatment (-40°C) [4].

The microbial spoilage of dry egg products is limited due to low water activity [16].

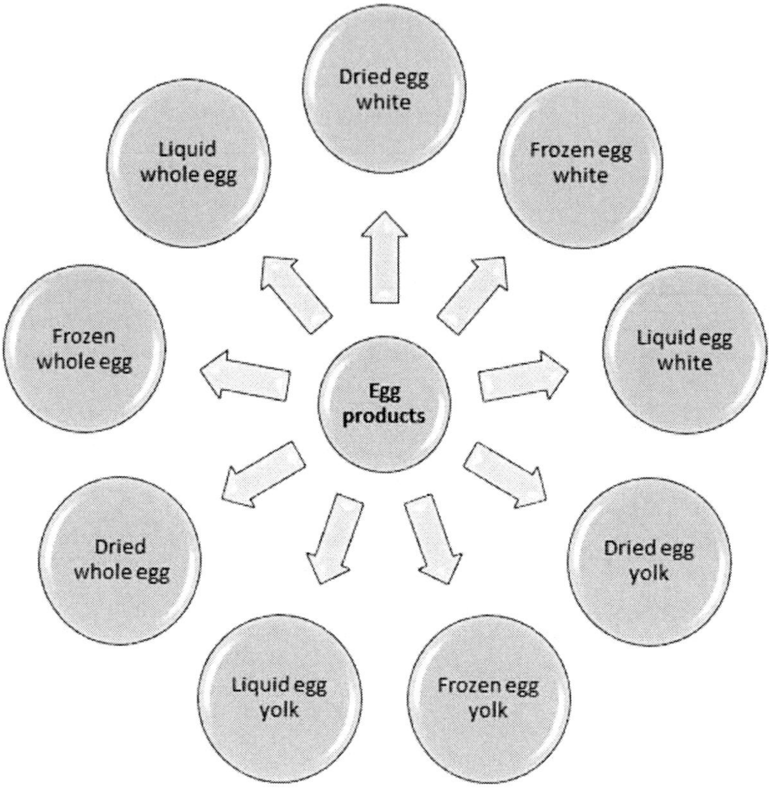

Figure 3.2. Schematic classification of egg products.

The pasteurization of egg products is difficult to apply because of the temperature sensitivity of the egg protein, but it is necessary for the inactivation of the microorganisms, especially the pathogens (*Salmonella* spp.) [4]. To avoid protein coagulation in the egg white, the thermal treatment recommended is 57.5°C / 13 min and 60°C / 5 min [16].

Pathogens in Egg and Egg Products

The egg is an important vector of foodborne pathogens that cause illness in humans, especially *Salmonella* spp. The egg contamination with *Salmonella* spp. may occur by vertical transmission and the serotypes involved are: *Salmonella* Heidelberg [23] and *Salmonella* Pullorum [24].

However, external contamination has been observed by *Salmonella* Typhimurium [25], *Salmonella* Infantis and *Salmonella* Hadar [26].

Other *Salmonella* serotypes isolated from egg shells are *Salmonella* Bareilly, *Salmonella* Derby, *Salmonella* Essen, *Salmonella* Montevideo, *Salmonella* Oranienburg. *Salmonella* Thompson, *Salmonella* Worthington, etc. [22].

In addition to *Salmonella*, other pathogens have also been isolated from egg and egg products, belonging to the *Campylobacter* spp., *Listeria* spp., *Staphylococcus* spp., *Aeromonas* spp and *Escherichia* spp. genera [16, 22].

EGGS QUALITY AND SAFETY

The egg for consumption must fulfill the following conditions to be accepted by consumers:

- *Shell quality*: the egg shell must be whole, clean and free of cracks;
- *Interior egg quality*: the white must be dense, transparent and the yolk slightly flattened and well defined by the white;

Frozen egg products must meet the following requirements:

- at the opening of the package, the contents should show a smooth surface;
- after defrosting, the liquid should be homogeneous, odorless, yellow in color for the yolk and white for the albumen;
- the pH should be 6.5-7 for egg mix (white and yolk) and 5.9 for yolk;

The powdered products (whole egg powder) must be yellow in color, homogeneous throughout the mass and not show stable agglomerations.

Both frozen products and egg powders are obtained only from hens' eggs from farms free from microbial contaminants transmissible to humans.

The eggs contaminants are usually enteric bacteria [27]. The egg content being suitable for the bacterial growth, especially *Salmonella* spp., the producers must ensure and apply control measures to prevent egg contamination, such as:

- the use of systems that minimize the possibility of cracking the eggs during handling;
- decontamination of the areas where birds are kept;
- vaccination of birds against pathogens;
- constant checking feed for the presence of potential pathogens.

After being collecting, eggs must be clean and kept in cold and clean places, and if possible, the contaminated eggs must be thermally treated [27].

Conclusion

Eggs are usually considered to be a relatively safe food, but that does not mean they cannot be contaminated with pathogens in various stages of the production process. Therefore, the manufacturers must ensure good sanitary practices for the production of safe eggs, and also the authorities must control and apply measures aimed at limitating pathogenic contamination.

References

[1] Li-Chan, Eunice, C. Y., and Hyun-Ock Kim. 2008. "Chemical Composition of Eggs". In *Egg Bioscience and Biotechnology*, edited

by Mine, Yoshinori, 1-95, John Wiley & Sons, Inc., Hoboken, New Jersey (Li-Chan and Kim 2008, 15).

[2] Baron, F. and S. Jan. 2011. "Egg and Egg Product Microbiology". In *Improving the Safety and Quality of Eggs and Egg Products*, vol. 1 Egg Chemistry, Production and Consumption, edited by Y. Nys, M. Bain, and F. Van Immerseel. Woodhead Publishing (Baron and Jan 2011, 331-333).

[3] Shebuski, Joseph R., and Timothy A. Freier. 2009. "Microbiological Spoilage of Eggs and Egg Products". In *Compendium of the Microbiological Spoilage of Foods and Beverages,* edited by Michael P. Doyl, 121-134, Springer, New York (Shebuski and Freier 2009, 122-124).

[4] Belitz, H.-D., Werner Grosch, Peter Schieberle. 2009. "Eggs". In *Food Chemistry*, edited by Belitz, H.-D., Werner Grosch, Peter Schieberle. 551-565, Springer-Verlag Berlin Heidelberg (Belitz et al. 2004, 552, 553, 556, 563, 565).

[5] Board, Ronald G. 2000. "Eggs and Egg Products". In *Microbiological Safety and Quality of Food*, edited by Barbara Lund, C. Anthony, Grahame Baird-Parker, and W. Gould, 590-619, Springer US (Board 2000, 591).

[6] Aygun, Ali. 2017. "The Eggshell Microbial Activity". In *Egg Innovations and Strategies for Improvements*, edited by Patricia Hester, 135-144, Academic Press (Aygun 2017, 137).

[7] Mine, Yoshinori., Cedric Oberle and Zeina Kassaify. 2003. "Eggshell Matrix Proteins as Defense Mechanism of Avian Eggs". *Journal of Agricultural Food Chemistry,* 51(1):249–253. doi: 10.1021/jf020597x (Mine et al. 2003, 249).

[8] Sparks, N. H. C. 2014. "Microbiology of Fresh Eggs". In *Encyclopedia of Food Microbiology,* edited by Carl A. Batt, 610-616, Academic press, London, UK (Sparks 2014, 612).

[9] Brady, D., S. Gaines, L. Fenelon, J. McPartlin and C. O'Farrelly, 2002. "A Lipoprotein-Derived Antimicrobial Factor from Hen-Egg Yolk is Active Against *Streptococcus* species". *Journal of Food*

Science, 67:3096–3103. doi: 10.1111/j.1365-2621.2002.tb08865.x (Brady et al. 2002, 3096).

[10] De Reu, K., M. Heyndrickx, K. Grijspreedt, B. Rodenburg, F. Tuyttens, M. Uyttendaele, J. Debevere and L. Herman. 2007. Estimation of the Vertical and Horizontal Bacterial Infection of Hen's Table Eggs. In *XVIII European symposium on the quality of poultry meat and XII European symposium on the quality of eggs and egg products* - Conference proceedings, 55-56, Prague, Czech Republic.

[11] Messens, W., K. Grijspeerdt and L. Herman. 2005. "Eggshell Penetration by *Salmonella*: A Review". *World's Poultry Science Journal,* 61(1):71-86 (Messens et al. 2005, 71).

[12] Raspoet, R., I. Gantois, R. Devloo, F. Pasmans, F. Haesebrouck, R. Ducatelle and F. Van Immerseel. 2011. "Internal Contamination of Eggs by *Salmonella* Enteritidis". In *Improving the safety and quality of eggs and egg products,* Volume 2: *Egg safety and nutritional quality*, edited by Van Immerseel Filip, Yves Nys, and Maureen Bain, 46-61, Woodhead Publishing Limited (Raspoet, et al. 2011, 48).

[13] De Reu, K., K. Grijspeerdt, W. Messens, M. Heyndrickx, M. Uyttendaele, J. Debevere and L. Herman. 2006. "Eggshell Factors Influencing Eggshell Penetration and Whole Egg Contamination by Different Bacteria, Including *Salmonella enteritidis". International Journal of Food Microbiology,* 112(3):253-260. doi: 10.1016/ j.ijfoodmicro.2006.04.011 (De Reu et al. 2006, 253).

[14] Techer, C., F. Baron and S. Jan. 2014. "Microbial Spoilage of Eggs and Egg Products". In *Encyclopedia of Food Microbiology,* edited by Batt, Carl A., 439-445, Academic press. London, UK (Techer 2014, 440).

[15] Broughton, J. Delves. 2014. "Microbiology of Egg Products". In *Encyclopedia of Food Microbiology,* edited by Carl A. Batt, 617-621, Academic press. London, UK (Broughton 2014, 617).

[16] Erkmen, Osman and T. Faruk Bozoglu. 2016. "Spoilage of Eggs and Egg Products". In *Food Microbiology: Principles into Practice,*

edited by Erkmen, Osman and T. Faruk Bozoglu, 296-300, John Wiley & Sons (Erkmen and Bozoglu 2016, 297-299).

[17] Corry, J. E. L. 2007. "Microbiological Analysis of Eggs and Egg Products". In *Microbiological Analysis of Red Meat, Poultry and Eggs*, edited by G. C. Mead, 183-201, Woodhead Publishing Limited, Cambridge, England (Corry 2007, 186-187).

[18] Board, R. G., H. S. Tranter, 1995. "The Microbiology of Eggs". In *Egg Science and Technology*, edited by W. J. Stadelman, O. J. Cotterill, 81–104, Food Products Press, New York, NY, United States (Board and Tranter 1995, 83).

[19] Hayes, P. R., and S. J. Forsythe. 1998. "Food Spoilage". In *Food Hygiene, Microbiology and HACCP*, edited by P. R. Hayes, and S. J. Forsythe, Springer Science and Business Media (Hayes and Forsythe 1998, 119).

[20] Ray Bibek and Arun Bhunia. 2013. "Spoilage of Specific Food Groups". In *Fundamental Food Microbiology*, edited by Ray Bibek and Arun Bhunia, 255-272, CRC Press (Ray and Bhunia 2013, 259).

[21] Hajmeer Maha N. and Daniel Y. C. Fung. 2006. "Infections with Other Bacteria". In *Foodborne Infections and Intoxications*, edited by Hans P. Riemann and Cliver, Dean O., 341-365, Academic Press (Hajmeer and Fung 2006, 357).

[22] Wareing, P., 2009. "Eggs". In *Microbiology Handbook - Meat Products*, edited by: R. Fernandes, 155–182. Leatherhead Publishing, United Kingdom (Wareing 2009, 159, 169).

[23] Gast, R. K., R. Guraya, J. Guard-Bouldin and P. S. Holt. 2007. "In Vitro Penetration of Egg Yolks by *Salmonella Enteritidis* and *Salmonella Heidelberg* Strains During Thirty-Six-Hour Ambient Temperature Storage". *Poultry Science,* 86(7):1431-1435. https://www.ncbi.nlm.nih.gov/pubmed/17575192 Accessed November 2, 2017 (Gast et al. 2007, 1431).

[24] Wigley, P., A. Berchieri, K. L. Page, A. L. Smith and P. A. Barrow. 2001. "*Salmonella enterica* serovar Pullorum persists in Splenic Macrophages and in the Reproductive Tract During Persistent, Disease-Free Carriage in Chickens". *Infection and Immunity*

69(12):7873-7879. doi: 10.1128/IAI.69.12.7873-7879.2001 (Wigley et al. 2001, 7873).

[25] Keller, Linda H., Charles E. Benson, Kristine Krotec and Robert J. Eckroade. 1995. "*Salmonella enteritidis* Colonization of the Reproductive Tract and Forming and Freshly Laid Eggs of Chickens". *Infection and Immunity* 63(7):2443-2449. https://www.ncbi.nlm.nih.gov/pmc/articles/PMC173326/ Accessed November 14, 2017 (Keller, et al. 1995. 2443).

[26] Okamura, M., T. Miyamoto, Y. Kamijima, H. Tani, K. Sasai and E. Baba. 2001. "Differences in Abilities to Colonize Reproductive Organs and to Contaminate Eggs in Intravaginally Inoculated Hens and in Vitro Adherences to Vaginal Explants Between *Salmonella Enteritidis* and Other *Salmonella Serovars*". *Avian Diseases,* 962-971. doi: 10.2307/1592875 (Okamura et al. 2001, 962).

[27] Chambers, James R., Khalid Zaheer, Humayoun Akhtar and El-Sayed, M. Abdel-Aal. 2017. "Chicken Eggs". In *Egg Innovations and Strategies for Improvements*, edited by Patricia Hester, 3-11. Academic Press (Chambers et al. 2017, 8).

In: Microbial Contamination of Food Product ISBN: 978-1-53613-524-4
Editors: G. Mocanu and O. Constantin

Chapter 4

MICROBIAL CONTAMINANTS OF FRUITS AND VEGETABLES

Oana Viorela Nistor*
Department of Food Science and Engineering,
Biotechnology and Aquaculture,
Faculty of Food Science and Engineering,
Dunărea de Jos University of Galați, Galați, România

ABSTRACT

Fruits and vegetables are products which are consumed fresh and processed. Fresh fruits and vegetables are the promoters of good health, but it is necessary to control foodborne pathogens, especially *Salmonella* spp., *Listeria monocytogenes*, *Escherichia coli*, and *Shigella* spp. There are many technologies to reduce or to eliminate the pathogen microorganisms present in food products.

The present study examines some of the challenges and limitations of non-thermal and thermal processing that have shown some microbial effects on vegetables.

* Corresponding Author: onistor@ugal.ro.

Keywords: food spoilage, foodborne infections, fresh-cuts, fermented vegetables

INTRODUCTION

An important and primary goal all over the world is food safety; international and national regulations are more and more restrictive, so there is a clear concern with the microbial contamination of fruits and vegetables.

It is well known that fruits and vegetables have gained an important place in human diet as a source of nutrients, vitamins and fibers. They also play vital roles in human health and well-being, especially due to their ability to prevent vitamin C and vitamin A deficiencies, which are known to reduce the risk of several diseases.

Fruits and vegetables are known to be important for the health and wellbeing of humans by regulating digestion processes, supplying slow-release sugar, reducing blood pressure, affecting the uptake and metabolism of fats, and possibly delaying the aging processes [1].

In the past decade, many benefits regarding the disappearance or the decrease of some illnesses (cardiovascular disease; cancer; diabetes; and osteoporosis) have been presumed to be supported by fruit and vegetable consumption.

The increased fruit and vegetable consumption imposes a significant concern with the high quality of nutritional, sensory and microbiological aspects of these foods [2].

Nowadays, in the contemporary lifestyles, fruits and vegetables are convenient meals. Various types of fruits and raw eaten vegetables and salads are also very popular due to their attributes, and the minimal time used to prepare them.

The microbial density for fruits and vegetables depends on the natural variability of the product, ranging from 10^3 to 10^7 CFU/g [3].

The dominant presence on fruits and vegetables is represented by lactic acid bacteria like *Lactobacillus* spp. and others (*Pseudomonas* spp.,

Erwinia spp., *Pantoea* spp., *Micrococcus* spp., *Flavobacterium*), and some Gram-positive spore formers (*Bacillus* spp., *Clostridium* spp.).

However, the most popular are yeasts (*Torulopsis* spp., *Saccharomyces* spp. and *Candida* spp.) and molds (*Alternaria* spp., *Penicillium* spp., *Fusarium* spp. and *Aspergillus* spp.); *Saccharomyces* spp. and *Candida* spp. are the dominant microorganisms for those fruits which have high sugar content [4].

To minimize the possible contamination risks it is necessary to identify the potential sources and to find the specific preventive measures. Microbial contamination can occur during any steps of the farm-to-table process, and this contamination can result from environmental, animal, human sources and technological applications [5].

The natural microbiota of raw fruits and vegetables is usually non-pathogenic for humans and may be present at the time of consumption [6]. During growth, harvest, transportation, handling and further processing, vegetables can be contaminated with pathogens from human, animal, or environmental sources [7]. Besides all these aspects, the most important concern remains the possible formation of foodborne pathogen biofilms on plant tissues [8]. The internal tissues are nutrient-rich, especially in vegetables, which have an almost neutral pH. Their structure mainly consists of the polysaccharides cellulose, hemicellulose, and pectin. The principal polymer stored is starch [9].

Most of the microorganisms which are capable of causing human disease may be found on raw produce. Furthermore, most of them are present as incidental contaminants from the soil, dust and the surroundings, or they can be introduced through poor production and handling practices.

The growth of microorganisms on fruits and vegetables depends on several factors such as: water, oxygen, pH, nutrients and the temperature.

Water is necessary to maintain the microorganisms' growth; dried legumes are not susceptible to spoilage. Fruits and vegetables are high-moisture foods that have a_w level of 0.98 or higher. *Oxygen* is particularly necessary to increase the number of aerobic microorganisms, found on the surface of the plant (e.g., molds).

The *pH values* for bacteria growth should not be very acidic, because most bacteria are neutrophiles, meaning that the growth is optimum at the neutral pH of 7. Most fruits, fruit juices, fermented fruits, vegetables, are low-acid foods (e.g., tomatoes have a pH of 4.1–4.4).

SPOILAGE MICROORGANISMS

It is quite important to know that spoilage is a process which can cause loss of the desired quality of fruits and vegetables.

Vegetables and fruits can be an excellent source for all types of microorganisms, which can be potential contaminants. The resulting spoilage of fruits and vegetables is characterized by a brown discoloration, necrosis, and loss of texture, exudation and/or production of off-odors or off-flavors [10].

Due to the growth conditions and handling in unprotected atmosphere, natural environments contribute to the level and diversity of microbial contamination [11].

Types of Spoilage Microorganisms

The most common types of spoilage are generally described by means of the term rot, which is associated with the terms specific to changes in the appearance, such as black rot, grey rot, pink rot, soft rot, stem-end rot.

The most important and commonly found spoilage- producing microorganisms in vegetables are molds belonging to the *Penicillium* spp., *Phytophthora* spp., *Alternaria* spp., *Botrytis* spp., *Aspergillus* spp. genera, and bacteria of the *Pseudomonas* spp., *Erwinia* spp., *Bacillus* spp., and *Clostridium* spp. species [12].

Due to the higher content of sugars and the pH, the spoilage microorganisms in fruits are molds (*Penicillium* spp., *Aspergillus* spp., *Alternaria* spp., *Botrytis* spp., *Rhizopus* spp.) and yeasts (*Saccharomyces* spp., *Candida* spp., *Torulopsis* spp., and *Hansenula* spp.), as well as aciduric bacteria (*Acetobacter* spp., *Gluconobacter* spp.).

Microbiological Spoilage of Fresh-Cut Fruits and Vegetables

Some fruits, like apples, strawberries, citrus fruits can be fermented by yeasts from the *Saccharomyces, Candida, Torulopsis*, and *Hansenula* genera. The souring of berries and figs can be associated with the bacterial spoilage which induces the growth of lactic acid and acetic acid.

Penicillium expansum is a blue mold specific to apples, which are postharvested. It is responsible not only for the damage, but also for the production of the carcinogenic mycotoxin named patulin. The rotten areas are soft, brown and watery. It can be recognized by the blue-green spots, which are white at first. The lesions are browned and it has a moldy smell [13].

Botrytis cinerea - it is named the grey mold and it is responsible for the rotting of grapes. It can also affect other fruits, such as strawberries [14].

Monilinia laxa or the brown rot can affect especially peaches, pears, apples and plums. It appears under the form of small brown squishy circles, which gradually spread over the surface of the entire fruit. This fungus can also affect the physiological parts of the plants, such as the flowers, leaves and stems of the tree. The ripened fruits are more susceptible to contamination than the raw ones. The warm and high humidity conditions accelerate the expansion of this fungus.

The *Xanthomonas* species are involved in fruit alterations that are associated with their pathogenic behavior, such as the presence of necrotic lesions.

Xanthomonas vesicatoria is a bacterium which affects sweet peppers and tomatoes. It can affect the leaves and fruits. The lesions appear small, irregular, like black or water-soaked spots. The spots are surrounded by a yellow halo [15].

Xanthomonas campestris could appear in the postharvest rot of plant crops (potato tubers, carrot, cucumber, cauliflower), beingone of the most important diseases in peaches, nectarines, Japanese plums, apricots, and almonds [16].

Erwinia carotovora is a bacterium which affects carrots, potatoes, tomatoes, leafy greens, squashes, onions and greens peppers. Therefore, it can affect any plant, leading to soft rot. Most plants or plant parts can resist invasion by the bacteria, unless the plant is already damaged. Furthermore, the high humidity and temperatures around 30 °C can favor the presence of *Erwinia carotovora.* Other factors which can facilitate its development are the plant enzymes: pectinases, cellulases, proteases, lipases, xylanases and nucleases [17].

The contamination of vegetables with *Pseudomonas* spp., Enterobacteriaceae, yeast and molds was also examined in a study developed by Stringer et al. [18] on broccoli florets and green beans.

In order to prevent or to reduce spoilage, fruits and fruit products are preserved by refrigeration, freezing, drying, and reducing water activity (a_W), vacuum packaging and heat treatment [12].

Apples remain the most widely sold fruit all over the world. Therefore, it is essential to ensure their microbiological safety during storage. The presence of pathogenic microorganisms on the apple surface during storage can promote the decay of apples and cause serious metabolic disorders in humans through the action of microorganism-derived toxins [19]. The dominating microscopic fungi before storage were *Penicillium* spp., *Alternaria* spp., *Botrytis* spp., *Aspergillus* spp. and *Cladosporium macrocarpum*. *Penicillium italicum* was the most abundant species of the *Penicillium* genus. Bacteria such as *Pseudomonas fluorescens* and *Bacillus cereus Penicillium italicum, Bacillus cereus,* and yeasts (e.g., *Candida curvata, Candida famata, Pichia etchellsii,* and *Pichia carsonii)* were also found on apples stored at refrigeration temperature.

Yeasts are the dominating microorganisms on apples after storage. A significant reduction of microorganisms from the apple surface was observed due to modified atmosphere with a low content of oxygen, as under these conditions aerobic microorganisms (e.g., molds) are not able to develop. So by using the controlled modified atmosphere microorganisms — aerobic bacteria — were reduced by 50% [20].

Microbiological Spoilage of Fermented Fruits and Vegetables

The term "fermented vegetables" refers to vegetables preserved by fermentation. Fermentation may be: low-acid (because of the naturally achieved and maintained pH of up to 4.6) which can be reached naturally due to the acids produced by the action of microorganisms or by adding different substances.

Acidified vegetables are products in which an acid is directly added to preserve any non-fermented vegetable with an initial pH above 4.6.

Pickled vegetables is a term applied specifically to any fermented or acidified vegetable using a solution that contains vinegar as the major acidifying agent [21]. The range of pickled vegetables is wide, depending on their area of origin. Therefore, olives, carrots, tomatoes, cucumbers, garlic, onion, cabbage, caperberries are some of the potential pickled vegetables.

The presence of the microorganisms involved in the fermentation process depends on the added salt concentration and other environmental factors.

The spoilage of fermented fruits and vegetables produced by bacteria is limited by the fruits' low pH which inhibits bacterial growth, therefore spoilage is usually determined by yeasts and molds, due to their acid-tolerant nature [22]. Vegetables and fruits have a low buffering capacity providing special conditions for spoilage fungi.

Microbial growth during the natural fermentation of brined vegetables may be seen as covering four stages: initiation, primary fermentation, secondary fermentation, and post-fermentation spoilage [23]. The presence of spoilage microorganisms depends on the water quality and salt concentration. For example, *Enterococcus faecalis, Leuconostoc mesenteroides, Pediococcus cerevisiae* are actively present during primary fermentation.

One common spoilage method is the formation of bloating (CO_2 forms gas pockets in cucumbers, olives). The presence of oxygen or nitrogen can reduce bloating formation, but it can induce the growth of yeasts and molds. Molds and yeasts may grow as a film on the surface of the liquid. A possible solution to prevent spoilage is pasteurization, although this method could not be used regularly without affecting the natural fermentation process.

Fermented cucumbers are susceptible to spoilage, therefore by isolation the pathogens identified were the *Pichia* and *Issatchenkia* yeast species, as well as 3 Gram-positive bacteria: *Lactobacillus buchneri, Clostridium* spp., and *Pediococcus thanolidurans*; and a Gram-negative bacterium *Enterobacter cloacae* [24, 25].

The spoilage microbiota can convert lactic acid to propionic acid. *Lactobacillus buchneri* is able to convert lactic acid to acetic acid and 1,2-propanediol under acidic conditions in silage and fermented cucumbers [26]. In the absence of fermentable sugar, *Lactobacillus plantarum* can metabolize lactic acid to acetic and formic acids. In olive fermentations, *Propionibacterium* spp. and *Clostridium* spp. have been considered as causative agents of spoilage, converting lactic acid to propionic and butyric acids [25].

Microbiological Spoilage of Fruit and Vegetable Juices

Juices are one of the most common and pleasant ways to consume fruits and vegetables. Juices are obtained by pressing fruits or vegetables, and if they are devoid of thermal treatment they are considered fresh juices.

The consumers' preference for fresh juices has increased because of the fresh flavor and the absence of preservatives. Fresh fruit juices are the liquid or the liquid with pulp extracted from fruits. Therefore, fresh juice is highly susceptible to spoilage more than the whole fruit, because in this case the fruits are unprotected by skin or cell walls.

Juices can be consumed as: fresh, thermally treated (concentrated, pasteurized, and sterilized), with or without pulp.

The main microorganisms present in juices are yeasts and molds and sometimes acid-tolerant bacteria. Microbiological stability in juices can be achieved by pasteurization or the use of chemical preservatives. Another important factor in the spoilage of juices is keeping juice in open packages for several days [27].

In fruit juices the most common microorganisms involved in spoilage are yeasts belonging to the *Pichia, Candida, Hansenula, Rhodotorula, Saccharomyces, Torulopsis, Trichosporon* and *Zygossacharomyces* genera that may be a result of the incorrect pasteurization of fruit juices and/or application of sanitation practices.

There are some specific yeast species that cause spoilage in citrus fruits, such as *Candida parapsilosis, Candida stellata, Saccharomyces cerevisiae, Torulaspora delbrueckii* and *Zygosaccharomyces rouxii*. Moreover, the *Rhodotorula rubra, Candida sake, Kloeckera apis* and *Candida lambica* are the most frequently encountered yeasts in apple, carrot, grape, grapefruit and orange juices. *Candida* spp., *Trichosporon mucoides, Kloeckera* sp., the yeast-like fungus *Cryptococcus neoformans* were observed in freshly squeezed juices of orange, lemon, grapefruit, and apple [28].

The pH values which allow the growth of yeasts and molds are □ 1.5 correlated to water activity values below 0.89. The growth of lactic acid bacteria is in the range of pH values 2.9– 3.5, while the pH values for the growth of the acetic acid bacteria are 3.0–4.5, and for the enteric bacteria 3.6–4.5; these values are higher than those for the growth of yeasts and molds. Most juices have a pH value within pH 3-4, and few samples have a pH value below pH 3, which means that mostly lactic acid bacteria, acetic acid bacteria and enteric bacteria will be able to grow within these juice samples [29].

Yeast spoilage in fruit juices is the most common and it is characterized by the formation of CO_2 and alcohol. Yeasts may also produce turbidity, flocculation, pellicles, and clumping. Due to their metabolism, they can produce pectinesterases which degrade pectin causing spoilage, organic acids, and acetaldehyde, which contribute to a fermented flavor [30].

Low acid vegetable juices are subject to spoilage by a variety of microorganisms which include Gram-negative bacteria like *Pseudomonas* spp., and various Enterobacteriaceae such as *Pectobacterium* spp., as well as Gram-positive bacterial spore-formers [9].

The concerns of fruit juice manufacturers regarding spoilage have revealed the presence of *Propionibacterium cyclohexanicum* and heat-resistant species of fungi such as *Byssochlamys fulva*, *Byssochlamys nivea*, and *Neosartorya fischeri,* as well as species of *Talaromyces* [31].

Molds in juice produce musty and stale off-flavors. The activity of pectinesterases leads to juice cloud loss. The most dominant molds recorded in fruit juices are *Penicillium* spp., *Cladosporium* spp., *Aspergillus niger, Aspergillus fumigatus, Botrytis* spp., and *Aureobasidium pullulans* [30].

For the thermally treated juices there are some specific differences - first of all the microflora is different depending on the type of thermal treatment. Juice pasteurization should assure a 5 log reduction of the most resistant microorganisms. Many studies found positive results of the thermal pasteurization HTLT (High Temperature Long Time - 80°C/holding time >30 s) during storage, for basil-bottle gourd juice [32], for grapefruit juice [33], spinach and sweet lime juices [34]. In the case of HTST treatment (High Temperature Short Time - 80°C/holding time ≤30 s), it provides productsafety. HTST treatments control the growth of *Lactobacillus plantarum* in orange juice [35], or the native microorganisms in the orange/sweet pepper juice blend [36] and mango nectar [37]. MTLT (Mild Temperature-Long Time - <80°C/holding times >30 s) can ensure 4.39 log reduction of aerobic plate count in pomegranate juice [38].

Therefore, fluid components are exposed to air and the microorganisms from the environment. Thus, unheated juice is subject to rapid microbial, enzymatic, chemical, and physical deterioration.

Sources of Spoilage Microorganisms

Considering the multitude of consumption possibilities of plant products, there is a wide variety of sources of contamination. The main sources of microorganisms in vegetables are soil, water, air, and other environmental sources, and can include some plant pathogens [12]. The presence and number of microorganisms differ, depending on the type of food, agronomic practices, geographical area of production, and weather conditions. Furthermore, harvest, transportation, handling and further processing can greatly influence the microbiota pattern [39].

Possible Action Mechanisms

Microbiological spoilage can generate different processes, such as production of enzymes or metabolites which results in visual and textural defects and off-odors. The composition of fruits and vegetables determines the type of spoilage. Therefore, carrots, bell peppers, and some others fresh-cut fruits, which are sugar-rich products, could be a good source for yeasts and lactic acid bacteria, which are responsible for off-odors, production of acids (lactic acid, acetic acid, malic acid, succinic acid, and pyruvic acid) [40].

Another reaction compound related to fruit and vegetable spoilage is ethanol. Ethanol production is given by bacteria and yeasts as Jacxsens et al. [41] demonstrated on a medium of mixed lettuce and agar. The fermentative metabolites such as ethanol can be produced as a fermentation reaction of fruits and vegetables when stored at very low O_2 concentrations or too high CO_2 concentrations.

The action mechanisms of molds can be sustained by temperature abuse, high initial mold contamination, which can appear as visual spoilage before the yeasts or bacteria can produce off-odors [42].

Different microorganisms produce pectinolytic enzymes that can influence textural changes in fresh-cut fruits and vegetables by degrading the pectin molecule in the middle lamella and the primary cell wall [43].

A part of the yeasts from juices are resistant to preservatives such as: *Zygosaccharomyces bailli, Candida krusei, Saccharomyces bisporus, Schizosaccharomyces pombe,* and *Pichia membranaefaciens*. The mechanism is based on the ability of cells to tolerate chronic intracellular pH drops by means of the phosphofructokinase enzyme [44]. Moreover, *Pichia membranaefaciens* is resistant to heat, moderate amount of salt,

SO_2, sorbic, benzoic and acetic acids, and it is considered as the target microorganism for the optimization of thermal pasteurization [45].

Pathogens in Fruits and Vegetables

Type of Pathogens

Types of Pathogens Associated with Fresh-Cut Fruits and Vegetables

Fresh-cut fruits and vegetables can be associated with pathogens as *Salmonella* spp., *Shigella, Escherichia coli* O157:H7, *Campylobacter* spp., *Listeria monocytogenes, Clostridium botulinum* and *Bacillus cereus*. All these pathogens have as sources the animals, water, soil or humans.

In a study conducted by Khiyami et al. [46] over foodborne pathogen contamination in minimally processed vegetable salads, *Enterococcus* spp., *Salmonella* spp., and *Shigella* spp were isolated.

Salmonella spp. and *Shigella* spp. are mesophilic feces-associated pathogens. The low pH of fruits limits their growth, but temperatures over 10-12°C can contribute to their growth. *Salmonella* spp. was involved in the contamination of some vegetables as lettuce, melon and tomatoes, cauliflower, sprouts, endive, spinach and mushrooms [47].

Imported food products from world regions where hygienic standards are insufficient have become a potential source of *Shigella* spp. able to contaminate foods. Therefore, refrigeration in proper conditions is the best preservation method to prevent an outgrowth of these mesophilic pathogens.

Clostridium botulinum was readily found in many vegetables, such as red peppers, carrots, onions, potatoes, parsley, spinach, garlic, cabbage, cultivated mushrooms [48].

Cabbage was found to be contaminated by *E. coli* and *Pseudomonas* spp. and also *Staphylococcus aureus,* which may be due to the poor hygiene during handling, processing and marketing the product, since *Staphylococcus aureus* is known to live in the nasal passage of healthy individuals [49].

Listeria monocytogenes was found in potatoes and carrots, cabbage, celery and spinach [50].

Aeromonas spp. was isolated in asparagus, broccoli, cauliflower, carrot, celery, cherry tomatoes, courgette, cucumber, lettuce, mushroom, pepper, turnip, escarole and watercress [51].

Types of Pathogens Associated with Pickled Fruits and Vegetables

Generally, if fermentation performed properly, the pathogens are destroyed during the process, thus making the final product safe for consumption. Pathogenic bacteria have not been reported in commercially fermented vegetable products, although the potential for the survival and growth of some pathogens in acidic environments has been investigated.

If *E. coli* O157:H7 has an extended survival, it creates the right environment conditions for the growth and development of *Enterobacter cloacae,* able to resist low pH and high salt concentrations, which were present in fermented olives [52].

Other fermented or acidified vegetable products were susceptible to the survival of *E. coli* O157 and *L. monocytogenes* during kimchi fermentation [53]. Also, in fermented cauliflower the survival of *L. monocytogenes* and *Salmonella typhimurium* was presented by Paramithiotis et al. [54].

Types of Pathogens Associated with Fruit and Vegetable Juices

The pasteurization of fruit juices at 90–95°C for 4–10 seconds should be effective against *E. coli* and *Salmonella*, but it is not effective against the ascospores of heat resistant fungi and heat resistant spore-forming bacteria [55].

The presence of *E. coli* O157:H7 in apple cider and *Salmonella* spp. in orange juice was identified in several studies, where good manufacturing practices were not applied. In apple juice with a pH of 3.3 and tomato juice with a pH of 4.0, *Shigella flexneri* and *Shigella sonnei* survived at 7°C for at least 14 days [56].

Sospedra et al. [57] reported the presence of *Salmonella* sp. and *Staphylococcus aureus* in orange juice extracted by squeezing machines used in restaurants. The pasteurized carrot juice was contaminated with *Clostridium botulinum* [58].

Vibrio cholorae was reported to have caused an outbreak in India by the consumption of unpasteurized sugarcane juice [59].

Juices have become frequent vehicle for transmitting pathogens such as enterohaemorrhagic *Escherichia coli O157, Salmonella*, and *Cryptosporidium* [60].

E. coli, *Staphylococcus* spp., *Moraxella* spp., *Klebsiella* spp., *Shigella* spp. and *Pseudomonas* spp. were found in different kinds of fruit juice samples (orange, pomegranate, grape, mango and litchi). *Bacillus* spp. and *Salmonella* spp. were present in jujube juice samples. Some other different microorganisms such as *Listeria* spp., *Clostridium* spp., *Acinetobacter* spp., *Brevibacterium* spp., *Acetobacter* spp., *Francisella* spp. have been identified in juice samples [29].

Mortuza [29] discovered that in packed mango juice the total bacterial count was 0.5×10^{3}- 1.25×10^{6} CFU/mL. The variation may be due to the unhygienic maintenance during juice preparation. Other researchers [61, 62] found that the total bacterial counts in juice samples ranged between 2.0×10^{6} to 1.0×10^{5} CFU/mL or 3.7×10^{8} CFU/mL in grape juice.

Types of Pathogens Associated with Canned Fruits and Vegetables

Clostridium botulinum was present in approximately 90% of all reported situations in the world related to the consumption of home-made preserves, especially canned vegetables [48].

In the case of jams, the increased quantity of sugar in the finished products, the evaporation effect of the thermal treatment and the a_w values are able to inhibit the germination of *Clostridium botulinum* spores.

The spore-formers *Bacillus* spp. and *Clostridium* spp., which are usually found in soil, can reach the surface of the plant, especially root vegetables.

The most dangerous to humans is *Clostridium perfringens,* which was isolated from the leaves of green plants like horse radish – 61%, and vegetables – up to 34% [63].

Bacillus subtilis and *Bacillus licheniformis* are the dominant raw materials contaminants and can survive as the residual microbiota of the product before sterilization. They were also detected in the residual microbiota of canned fruits and vegetables. The presence of these microorganisms in canned products may lead to reduced immunity and cause various diseases in humans [63].

Sources of Pathogens

Pathogenic bacteria such as *Clostridium botulinum, Bacillus cereus* and *Listeria monocytogenes* are normal inhabitants of soils, whereas *Salmonella, Shigella, Escherichia coli* and *Campylobacter* can reside in the intestinal tracts of animals, including humans, and are more likely to contaminate raw fruits and vegetables through contact with feces, sewage, untreated irrigation water or surface water [64]. Another possible sources are the handling of products, the harvesting and processing equipment, and transportation. The natural growth of fruits and vegetables is a possible source of pathogens, mainly provided by humans.

Soil Amendments

Another potential source of contamination can be generated by the use of natural soil amendments for the organic fruits and vegetables. In this case it is important to manage the quantity and to apply some special measures in order to reduce the number of pathogens. The good manufacturing practices are difficult to integrate in the development and implementation of microbial risk prevention and reduction programs at the farm. Therefore, the only chance to reduce the number of pathogenic microorganisms and to prevent foodborne illnesses is to wash or cook the food thoroughly, or to apply an adequate thermal treatment to the organic products.

Irrigation Water

The quality of the irrigation water depends on its source — whether it is surface water, groundwater, rainwater or drinkable water. Ground water seems to be safer than surface water, which can be contaminated by the wastewaters, stormwaters and agricultural run-off.

More than the presence of pathogens in water, the irrigation technique can cause the contamination of fruits and vegetables. The irrigation techniques are spraying (which increases the risks of contamination), dripping and flooding. The hydroponic water systems seem to be more useful than the other types which consist of sewage tanks, but they cannot decrease the risks of contamination, if the water and the tanks are not hygienically treated [65].

Several researchers isolated in a study conducted in crops (lettuce, carrots and amaranthus) bacteria such as *Salmonella* spp., *Vibrio* spp. and *E. coli* as a result of irrigation with water [66].

In the case of spinach leaves, it was found that the internalization of *E. coli* O157:H7 was caused not by soil contamination, but by the water dropped on the leaves.

Intestinal Tract of Humans and Animals

Several specific bacteria species are usually found in the intestinal tract of humans and animals, such as: *Bacteroides* spp., *Escherichia* spp., *Lactobacillus* spp., *Proteus, Salmonella* spp., *Staphylococcus* spp., and *Streptococcus* spp.

The most important natural inhabitant of the intestinal tract of humans and animals is the *Escherichia* genus. From the intestinal tracts of animals, microorganisms find their way directly to the soil and water and from here they may transfer onto plants, dust, utensils, and so on [67].

The contamination of fresh juices is a consequence of the poor washing of fruits, the use of unclean water for dilution, the dressing with ice, prolonged preservation without refrigeration, unhygienic surroundings and airborne dust, which can all act as sources of contamination. Such juices have been found to be potential sources of bacterial pathogens; notably *Escherichia coli* 0157:H7, *Salmonella* spp., *Shigella* spp., and *Staphylococcus aureus* [68].

Survival of Pathogens in Fruits and Vegetables

If pathogens will survive in the gastrointestinal tract, they can produce symptoms as nausea, vomiting, abdominal cramps and diarrhea, which are the main symptoms of foodborne diseases. Foodborne diseases (FBD) are divided into three groups as follows: intoxication, infection and food poisoning.

The survival and growth patterns of pathogens were dependent on pH, vegetable type, package atmosphere, storage temperature, and bacterial strain.

Lettuce, sprouted seeds, melon and tomatoes could be involved in *Salmonella* infection. *Salmonella* spp. are often isolated from produce sampled in routine surveys, including lettuce, cauliflower, sprouts, mustard cress, endive and spinach.

Shigella spp. is a very resistant microorganism, which survived for several days at both ambient (22°C) and refrigeration temperatures (5 and 10°C) when inoculated on salads and vegetables (carrots, coleslaw, radishes, broccoli, cauliflower, lettuce, and celery). The number of days of survival was almost extended to 11 days for coleslaw, carrot salad, and potato salad [39]. Bell peppers were not a good source of *Shigella* survival due to their lower pH.

Campylobacter spp. could survive at refrigeration temperatures for extended periods. The *Campylobacter* survival rate is longest at low oxygen concentrations, and modified atmosphere packaging may prolong the shelf life of this pathogen [69].

The survival and growth of *L. monocytogenes* on modified atmosphere packaged fresh-cut vegetables depended on the type of vegetables and the storage temperature [70].

Listeria spp. are versatile microorganisms due to their ability to continue to multiply at cold temperatures and survive in frozen foods. For the ready to eat food that is contaminated with *Listeria* spp., the bacteria can multiply even if refrigerated correctly at or below 5°C [71].

Many studies on *Escherichia coli* (largely *E. coli* O157:H7) and *Salmonella* spread on different parts of the plants (seeds, leaves, roots or directly into soil) demonstrated their prolonged survival and multiplication, for 177 and 231 days respectively. On lettuce, the water contaminated with *E. coli* O157:H7 induces the presence of pathogens for 30 days [72].

Clostridium botulinum remains viable for long periods of time even when environmental conditions are not favorable to their development. Sometimes the intrinsic factors (pH, water activity, redox potential values, microbial antagonism, preservatives) and extrinsic variables (temperature, retention time) could be insufficient to inactivate this microorganism, so the foods need to be thermally treated for a sufficient length of time [48].

A range of fresh fruit and vegetable products have been implicated in *Salmonella* infection, most commonly lettuce, sprouted seeds, melon and tomatoes. *Salmonella* spp. are often isolated from produce sampled in routine surveys, including lettuce, cauliflower, sprouts, mustard cress, endive and spinach.

Bacteriologically safe fruits and vegetables are essential to maximize the health benefits of the consumers. The proper washing of fruits and vegetables is essential for decontamination.

In conclusion, washing the vegetables with enough drinkable water at the proper time may have a significant role in reducing contaminations [73].

Methods to Reduce Microbial Contamination

All over the world many methods have been developed to reduce the number of microorganisms on whole and fresh-cut fruits and vegetables. Despite these efforts, the fact that foodborne illnesses continue to exist points to some disadvantages of these methods, which depend on the type of produce, work protocol, and other variables. Therefore, the prevention of microbial contamination continues to be performed by washing or sanitizing fruits and vegetables.

The efficiency of the method to reduce the microbial population depends on the type of treatment, the attachment of cells to produce surfaces, the internalization of microorganisms, the exposure time and concentration of sanitizer, pH, and temperature [74].

The prevention of microbial growth and the thermal treatment remain the most efficient methods to reduce microbial risks. Prevention can be assured by controlling the temperature of fruits and vegetables throughout the refrigeration chain (transporting vehicles, retail outlets, and domestic settings).

The preservation of fruits and vegetables can be done by chilling, freezing and at ambient temperature. Many fruits and vegetables are stored at refrigeration temperature (bellow 8°C) for 1-6 weeks. Even the storage at chill temperatures will reduce the growth rates of microorganisms, but many spoilage and/or pathogenic microorganisms are able to grow at refrigeration temperatures [75]. The fruits and vegetables can be stored at -18°C for 6 months to 2 years. At these low temperatures, the growth of microorganisms is not supported.

Dried fruits and vegetables which are subjected to a heat treatment (concentration, sterilization, drying) to destroy microbial contaminants, can be stored at room temperature.

Another method to prevent microbial contamination is the choice of the right packaging type/method.

In order to decrease microbial contamination, it is necessary to apply a type of thermal treatment such as: pasteurization to inactivate vegetative microorganisms and psychrotrophic or acid-tolerant spore formers, and sterilization. Another class of treatments are the minimal processing technologies such as high (hydrostatic) pressure processing (HPP), ultraviolet (UV) processing, and gamma irradiation.

The technology of HPP was successfully used for apricot and carrot juices [76] to improve the microbiological quality.

Drying is one of the oldest preservation methods and the most efficient because of the decrease of the moisture level to 10-15%. At this level, the microorganisms cannot thrive and the enzymes become inactive.

The prevention of the microbiological spoilage of foods can be achieved by the addition of antimicrobial food preservatives in order to inhibit the growth of foodborne pathogens [77]. There are many substances used to prevent the spoilage or the contamination with pathogens. Depending on the different mechanisms of action and on their chemical structure, the most commonly used preservatives include salt, sugar, dextrose, spices, vinegar, ascorbic acid, benzoic acid and its salt, SO_2 and the salts of sulphuric acid, nitrates, sorbic acid and its salts, propionic acid and its salts, lactic acid and its salts [78].

Their efficiency depends on the food composition and environmental factors, as a_w or the presence of other substances which can modify the antimicrobial action of preservatives [79].

Sodium benzoate and its similar compounds could negatively influence the quality of the preserved vegetables like olives or fruits, to which they can induce browning. As a consequence, the use of sodium metabisulphite, zinc chloride, or natamicin is a more suitable method in this particular case.

EU Regulations

The food safety and process hygiene criteria for ready to eat sprouted seeds and fresh-cut fruits and vegetables, as specified under EC/2073/2005 mentioned some limits, especially for the fresh-cut fruits and vegetables, as follows: *Salmonella* shall be absent in a sample of 25 g of pre-cut fruits and vegetables and fresh juices, while for *E. coli* the limits are defined for 100-1000 CFU/g for the same samples [80].

Conclusion

Fruits and vegetables will always be consumed and definitely be the core of food production aimed at supporting a constantly growing population. If certain hygienic rules are properly observed, the consumption of fruits and vegetables can be safe.

Moreover, the use of traditional fermentations is in line with the demand for natural, healthier food products. Fermented vegetables are considered microbially safe and hygienic products.

REFERENCES

[1] www.freshfel.org/20.11.2017/6:44.

[2] Chemat, Farid, Zill-e-huma, and Khan, M. K. 2011. "Applications of Ultrasound in Food Technology: Processing, Preservation and Extraction". *Ultrasonic Sonochemistry* 18(4):813-35. doi: 10.1016/j.ultsonch.2010.11.023 (Chemat and Khan 2011, 817).

[3] Badosa, Esther, R. Trias, D. Parés, M. Pla, and E. Montesinos, 2008. "Microbiological Quality of Fresh Fruit and Vegetable Products in Catalonia (Spain) Using Normalised Platecounting Methods and Real Time Polymerase Chain Reaction (QPCR)". *Journal of the Science of Food and Agriculture* 88: 605-11. doi: 10.1002/jsfa.3124 (Badosa et al. 2008, 608).

[4] Trias Mansilla, Rosalia. 2008. *"Lactic Acid Bacteria as Bioprotective Agents Against Foodborne Pathogens and Spoilage Microorganisms in Fresh Fruits and Vegetables".* Doctoral Thesis, Universitat de Girona. ISBN: 978-84-691-5683-84 (Trias 2008, 2).

[5] Serkan, Kemal Buyukunal, Issa Ghassan, Aksu Filiz, and Vural Aydin. 2015. "Microbiological Quality of Fresh Vegetables and Fruits Collected from Supermarkets in Istanbul, Turkey". *Journal of Food and Nutrition Sciences* 3(4):152-59. doi: 10.11648/j.jfns.20150304.13 (Serkan et al. 2015, 152).

[6] Food & Drug Administration, H. H. S. 2008. "Irradiation in the Production, Processing and Handling of Food". *Final rule. Federal Register*, 73(164): 49593–603 (FDA 2008, 49597).

[7] Sánchez, G., P. Elizaquível, and R. Aznar. 2012. "A Single Method for Recovery and Concentration of Enteric Viruses and Bacteria from Fresh-Cut Vegetables". *International Journal of Food Microbiology*

152(1–2):9–13. doi: 10.1016/j.ijfoodmicro.2011.10.002 (Sánchez et al. 2012, 9).

[8] Critzer, Faith J., and Doyle, M. P. 2010. "Microbial Ecology of Foodborne Pathogens Associated with Produce". *Current Opinion in Biotechnology* 21(2): 125–30. doi: 10.1016/j.copbio.2010.01.006 (Critzer and Doyle 2010, 128).

[9] Barth, Margaret, T. Hankinson, H. Zhuang, and F. Breidt. 2009. "Microbiological Spoilage of Fruits and Vegetables". In *Compendium in the Microbiological Spoilage of Foods and Beverages*, edited by W. Sperber, and M. Doyle, 135-83. Springer Science-Business Media, LLC, London (Barth et al. 2009, 167).

[10] Ponce, A. G., S. I. Roura, C. E. del Valle, and R. Fritz. 2002. "Characterization of Native Microbial Population cf Swiss Chard (Beta vulgaris, type cicla)". *LWT-Food Science and Technology* 35(4):331–37. doi: 10.1006/fstl.2001.0879 (Ponce et al. 2002, 333).

[11] Leff, Jonathan W., and Noah Fierer. 2013. "Bacterial Communities Associated with the Surface of Fresh Fruits and Vegetables". *PLOS One* 8(3): 59304-59310. doi: 10.1371/journal.pone.0059310 (Leff and Fierer 2013, 59304).

[12] Dilbaghi, Neeraj, and S. Sharma. 2007. *"Food and Industrial Microbiology: Food spoilage, Food Infections and Intoxications Caused by Microorganisms and Methods for Their Detection".* Available at: http://nsdl.niscair.res.in/bitstream/123456789/386/2/ FoodSpoilage.pdf (Dilbaghi and Sharma 2007, 12, 15, 16).

[13] Rundberget, Thomas, I. Skaar, and A. Flåøyen. 2004. "The Presence of Penicillium and Penicillium mycotoxins in Food Wastes". *International Journal of Food Microbiology* 90:181-88. doi: 10.1016/S0168-1605(03)00291-5 (Rundberget et al. 2004, 184).

[14] Jacometti, Marco. A., S. D. Wratten, and M. Walter. 2006. "Understorey Management Increases Grape Quality, Yield and Resistance to *Botrytis cinerea*". *Agriculture, Ecosystems & Environment* 122:349-56. doi: 10.1016/j.agee.2007.01.021 (Jacometti et al. 2006, 352).

[15] Gürlebeck, Doreen, F. Thieme, and U. Bonas. 2006. "Type III Effector Proteins from the Plant Pathogen Xanthomonas and their Role in the Interaction with the host plant". *Journal of Plant Physiology* 163:233-55. doi: 10.1016/j.jplph.2005.11.011 (Gürlebeck 2006, 239, 248).
[16] Boudon, Sylvain, C. Manceau, and J. L. Notteghem. 2005. "Structure and Origin of Xanthomonas arboricola pv. pruni Populations Causing Bacterial Spot of Stone Fruit Trees in Western Europe". *Phytopathology* 95:1081-88. doi: 10.1094/PHYTO-95-1081 (Boudon et al. 2005, 1081).
[17] Mijan Hossain, M., S. Shibata, S. I. Aizawa, and S. Tsuyumu. 2005. "Motility is an Important Determinant for Pathogenesis of *Erwinia carotovora* subsp. *carotovora*". *Physiological and Molecular Plant Pathology* 66:134-43. doi: 10.1016/j.pmpp.2005.06.001 (Mijan et al. 2005, 139).
[18] Stringer, S. C., J. Plowman, and M. W. Peck. 2007. "The Microbiological Quality of Hot Water-Washed Broccoli Florets and Cut Green Beans". *Journal of Applied Microbiology* 102:41–50. doi: 10.1111/j.1365-2672.2006.03065.x (Stringer et al. 2007, 41).
[19] Beretta, Barbara, A. Gaiaschi, C. L. Galli, and P. Restani. 2000. "Patulin in Apple Based Foods: Occurrence and Safety Evaluation". *Food Additives & Contaminants* 17: 399–406. doi: 10.1080/026520300404815 (Beretta et al. 2000, 402).
[20] Juhneviča, Karina, Skudra, Gita, and Skudra, Liga. 2011. "Evaluation of Microbiological Contamination of Apple Fruit Stored in a Modified Atmosphere". *Environmental and Experimental Biology* 9:53–59. Available at: http://eeb.lu.lv/EEB/201108/EEB_9_Juhnevica.pdf (Juhneviča et al. 2011, 58).
[21] Perez-Dıaz, Ilenys. M., F. Breidt, R. W. Buescher, F. N. Arroyo-Lopez, R. Jimenez-Dıaz, A. Garrido Fernandez, J. Bautista Gallego, S. S. Yoon, and S. D. Johanningsmeier. 2013. "Fermented and Acidified Vegetables". In *Compendium of Methods for the Microbiological Examination of Foods*, edited by APHA Press (Perez-Dıaz et al. 2013, 1).

[22] Khan, S. H., Muhammad Faqir, M. Idrees, M. Shafique, Imtiaz Hussain, and M. H. Farooq. 2005. "Some Studies on Spoilage Fungi of Pickles". *Journal of Agriculture & Social Sciences* 1(1):1813–2235. Available at: http://www.fspublishers.org/published_papers/82411.pdf (Khan et al. 2005, 15).

[23] Plengvidhya, Vethachai, F. Breidt, Z. Lu, and H. P. Fleming. 2007. "DNA Fingerprinting of Lactic Acid Bacteria in Sauerkraut Fermentations". *Applied and Environmental Microbiology* 73:7697-702. PMID 17921264 [PubMed - indexed for MEDLINE] (Plengvidhya et al. 2007, 7698).

[24] Franco, Wendy, I. M. Perez-Dıaz. 2012. "Role of Selected Oxidative Yeasts and Bacteria in Cucumber Secondary Fermentation Associated with Spoilage of the Fermented Fruit". *Food Microbiology* 32:338–44. doi: 10.1016/j.fm.2012.07.013 (Franco and Perez-Diaz 2012, 340).

[25] Franco, Wendy, I. M. Perez-Dıaz, S. D. Johanningsmeier, R. F. McFeeters. 2012. "Characteristics of Spoilage-Associated Secondary Cucumber Fermentation". *Applied and Environmental Microbiology* 78(4):1273–84. doi: 10.1128/AEM.06605-11 (Franco et al. 2012, 1277).

[26] Johanningsmeier, Suzanne D., W. Franco, I. M. Perez-Dıaz, R. F. McFeeters. 2012. "Influence of Sodium Chloride, pH, and Lactic Acid Bacteria on Anaerobic Lactic Acid Utilization During Fermented Cucumber Spoilage". *Journal of Food Science* 77(7):397–804. doi: 10.1111/j.1750-3841.2012.02780.x (Johanningsmeier et al. 2012, 398).

[27] Subramaniam, Persis, and Peter Wareing. 2016. "The Stability and Shelf Life of Food. Second Edition". *Series in Food Science, Technology and Nutrition,* edited by Woodhead Publishing 1-595. Amsterdam (Subramaniam and Wareing 2016, 57, 359).

[28] http://shodhganga.inflibnet.ac.in/bitstream/10603/117765/7/07_chapter%202.pdf/8.12.2017/6.15.

[29] Mortuza, Tasnim. 2016. *"Isolation and Identification of Microbes from Various Fruit Juices Made and Sold for Immediate*

Consumption at Home and in the Market of Dhaka City". Dissertation. Dhaka, Bangladesh (Mortuza 2016, 30).

[30] Lawlor, Kathleen A., J. D. Schuman, P. G. Simpson, and P. J. Taormina. 2009. "Microbiological Spoilage of Beverages". In *Compendium of the Microbiological Spoilage of Foods and Beverages, Food Microbiology and Food Safety*, edited by W. H. Sperber and M. P. Doyle. 1-367. Springer Science and Business Media, New York, USA (Lawlor et al. 2009, 164).

[31] Walker, Michelle, and C. A. Phillips. 2008. "The Effect of Preservatives on *Alicyclobacillus acidoterrestris* and *Propionibacterium cyclohexanicum* in Fruit Juice". *Food Control*. 19(10): 974–81. doi: 10.1016/j.foodcont.2007.10.003 (Walker and Phillips 2008, 976).

[32] Majumdar, T. K., D. D. Wadikar, C. R. Vasudish, K. S. Premavalli, and A. S. Bawa. 2011. "Effect of Storage and Physio-Chemical, Microbiological and Sensory Quality of Bottled Gourd-Basil Leaves Juices". *American Journal of Food Technology* 6:226–34. doi: 10.3923/ajft.2011.226.234 (Majumdar et al. 2011, 227).

[33] Uckoo Ram M., G. K. Jayaprakasha, J. A. Somerville, V. M. Balasubramaniam M. Pinarte, and B. S. Patil. 2013. "High Pressure Processing Controls Microbial Growth and Minimally Alters the Levels of Health Promoting Compounds in Grapefruit (Citrus paradisi Macfad) juice". *Innovative Food Science and Emerging Technology* 18:7–14. doi: 10.1016/j.ifset.2012.11.010 (Uckoo et al. 2013, 10).

[34] Khandpur, Paramjeet, and P. R. Gogate. 2015. "Effect of Novel Ultrasound Based Processing on the Nutrition Quality of Different Fruit and Vegetable Juices". *Ultrasonics Sonochemistry* 27:125–36. doi: 10.1016/j.ultsonch.2015.05.008 (Khandpur and Gogate 2015, 128).

[35] Zulueta, Ana, F. J. Barba, M. J. Esteve, and A. Frigola 2013. "Changes in Quality and Nutritional Parameters during Refrigerated Storage of an Orange Juice–Milk Beverage Treated by Equivalent Thermal and Non-Thermal Processes for Mild Pasteurization". *Food*

Bioprocess Technology 6:2018–30. doi: 10.1007/s11947-012-0858-x (Zulueta et al. 2013, 2022).

[36] Xu, Zhenzhen, T. Lin, Y. Wang, and X. Liao. 2015. "Quality Assurance in pepper and Orange Juice Blend Treated by High Pressure Processing and high Temperatureshort Time Treatment". *Innovative Food Science and Emerging Technology* 31:28–36. doi: 10.1016/j.ifset.2015.08.001 (Xu et al. 2015, 31).

[37] Liu, Fengxia, R. Li, Y. Wang, X. Bi, and X. Liao. 2014. "Effects of high Hydrostatic Pressure and High-Temperature Short-Time on Mango Nectars: Changes in Microorganisms, Acid Invertase, 5-hydroxymethylfurfural, Sugars, Viscosity, and Cloud". *Innovative Food Science and Emerging Technology* 22:22-30. doi: 10.1016/j.ifset.2013.11.014 (Liu et al. 2014, 26).

[38] Mena, Pedro, N. Marti, D. Saura, M. Valero, C. Garcia-Viguera. 2013. "Combinatory Effect of Thermal Treatment and Blending on the Quality of Pomegranate Juices". *Food and Bioprocess Technology* 6:3186–99. doi: 10.1007/s11947-012-0961-z (Mena et al. 2013, 3190).

[39] Olaimat, Amin N., and R. A. Holley. 2012. "Factors Influencing the Microbial Safety of Fresh Produce: A Review". *Food Microbiology* 32(1):1-19. doi: 10.1016/j.fm.2012.04.016 (Olaimat and Holley 2012, 2).

[40] Ragaert, Peter, Jacxsens, Liesbeth, Vandekinderen, Isabelle, Baert, Leen, and Devlieghere, Frank. 2011. "Microbiological and Safety Aspects of Fresh-Cut Fruits and Vegetables". In *Advances in Fresh-Cut Fruits and Vegetables Processing*, Edited by: Olga Martín-Belloso and Robert Soliva-Fortuny, CRC Press, Taylor&Francis Group (Ragaert et al. 2011, 56).

[41] Jacxsens, L., F. Devlieghere, P. Ragaert, E. Vanneste, and J. Debevere. 2003. "Relation between Microbiological Quality, Metabolite Production and Sensory Quality of Equilibrium Modified Atmosphere Packaged Fresh-Cut Produce". *International Journal of Food Microbiology* 83:263–80. doi: 10.1016/S0168-1605(02)00376-8 (Jacxsens et al. 2003, 268).

[42] Nielsen Tim, and A. Leufven. 2008. "The Effect of Modified Atmosphere Packaging on the Quality of Honeoye and Korona Strawberries". *Food Chemistry* 107:1053–63. doi: 10.1016/ j.foodchem.2007.09.025 (Nielsen and Leufven 2008, 59).

[43] Artes, Francisco, P. A. Gomez, and F. Artes-Hernandez. 2007. "Physical, Physiological and Microbial Deterioration of Minimally Fresh Processed Fruits and Vegetables". *International Food Science and Technology* 13:177–88. doi: 10.1177/1082013207079610 (Artes et al. 2007, 181).

[44] ICMSF. 2005. "Soft Drinks, Fruit Juices, Concentrates and Food Preserves". *Microorganisms in Foods 6: Microbial Ecology of Food Commodity*, edited by Kluwer Academic Publisher (ICMSF 2005, 223).

[45] Bevilacqua, Antonio, M. R. Corbo, D. Campaniello, D'Amato, M. Gallo, B. Speranza, and M. Sinigaglia. 2011. "Shelf Life Prolongation of Fruit Juices Through Essential Oils and Homogenization: A Review". *Science against Microbial Pathogens: Communicating Current Research and Technological Advances*, 1156–66. https://pdfs.semanticscholar.org/eb5a/5fc68aee47d230138b 638ae2e0bfcadf07cc.pdf (Bevilacqua et al. 2011, 1158).

[46] Khiyami, Mohammad, N. Al-Faris, B. Busaeed, and H. Sher. 2011. "Food Borne Pathogen Contamination in Minimally Processed Vegetable Salads in Riyadh, Saudi Arabia". *Journal of Medical Plants Research* 5(3):444-51. Available online at http://www. academicjournals.org/JMPR (Khiyami et al. 2011, 444).

[47] Doran, Geraldine, F. Sheridan, N. DeLappe, C. O'Hare, W. Anderson, Corbett-G. Feeney, and M. Cormican. 2005. "Salmonella enterica serovar Kedougou Contamination of Commercially Grown Mushrooms". *Diagnostic Microbiology and Infectious Disease* 51:73–76. doi: 10.1016/j.diagmicrobio.2004.08.006 (Doran et al. 2005, 74).

[48] Bhagat, Arpan, G. Caruso, M. Micali, and S. Parisi, 2016. *"Foods of non-Animal Origin: Chemistry, Technology, Inspection Procedures".*

edited by Springer International Publishing AG, Cham, Switzerland (Bhagat et al. 2016, 28).

[49] Ogbonna Okonkwo I., A. H. Ahmed, H. S. Waba, S. H. Bello, and O. O. Akinmusere. 2010. "Bacteriological Quality of Fruits and Vegetables Sold in Maiduguri, and their Effects of Some Antimicrobial Agents on the bacterial load". *Nigerian Journal of Experimental and Applied Biology* 11(1):63- 68 (Ogbonna et al. 2010, 65).

[50] Crepet Amelie, I. Albert, C. Dervin, and F. Carlin. 2007. "Estimation of Microbial Contamination of Food from Prevalence and Concentration Data: Application to *Listeria monocytogenes* in Fresh Vegetables". *Applied and Environmental Microbiology* 73:250–58 (Crepet et al. 2007, 252).

[51] Isonhood, Jamie H., and M. Drake. 2002. "Aeromonas species in Food". *Journal of Food Protection* 65:575-82. Available at: http://jfoodprotection.org/doi/pdf/10.4315/0362-028X-65.3.575?code=fopr-site (Isonhood and Drake 2002, 579).

[52] Pereira, Ana P., J. A. Pereira, A. Bento, and M. L. Estevinho. 2008. "Microbiological Characterization of Table Olives Commercialized in Portugal in Respect to Safety Aspects". *Food and Chemical Toxicology* 46:2845-2902. doi: 10.1016/j.fct.2008.05.033 (Pereira et al. 2008, 2876).

[53] Cho, Ga Y., Lee, M. H., and Choi, C. 2011. "Survival of *Escherichia coli* O157:H7 and *Listeria monocytogenes* During Kimchi Fermentation Supplemented with Raw Pork Meat". *Food Control* 22:1253–60. doi: 10.1016/j.foodcont.2011.01.026 (Cho et al. 2011, 1256).

[54] Paramithiotis, Spiros, A. I. Doulgeraki, I. Tsilikidis, G. J. E. Nychas, E. H. Drosinos. 2012. "Fate of *Listeria monocytogenes* and Salmonella typhimurium during spontaneous cauliflower fermentation". *Food Control* 27:178–83. doi: 10.1016/j.foodcont.2012.03.022 (Paramithiotis et al. 2012, 179).

[55] Danyluk, Michelle D., L. M. Friedrich, C. Jouquand, R. Goodrich-Schneider, M. E. Parish, and R. Rouseff. 2011. "Prevalence,

Concentration, Spoilage, And Mitigation of *Alicyclobacillus* spp. In Tropical and Subtropical Fruit Juice Concentrates". *Food Microbiology* 28(3):472–7. doi: 10.1016/j.fm.2010.10.008 (Danyluk et al. 2011, 474).

[56] Van, Opstal, I., C. F. Bagamboula, T. Theys, S. C. M. Vanmuysen, and C. W. Michiels. 2006. "Inactivation of *Escherichia coli* and *Shigella* in acidic Fruit and Vegetable Juices by Peroxidase Systems". *Journal of Applied Microbiology* 101(1):242–50. doi: 10.1111/j.1365-2672.2006.03002.x (Van Opstal et al. 2006, 242).

[57] Sospedra, I., J. Rubert, J. M. Soriano, and J. Manes. 2012. "Incidence of Microorganisms from Fresh Orange Juice Processed by Squeezing Machines". *Food Control* 23(1):282–85. doi: 10.1016/j.foodcont. 2011.06.025 (Sospedra et al. 2012, 282).

[58] Marshall, K. M., L. Nowaczyk, B. H. Raphael, G. E. Skinner, N. Rukma Reddy. 2014. "Identification and genetic characterization of *Clostridium botulinum* serotype A strains from commercially pasteurized carrot juice". *Food Microbiology* 44:149-55. doi: 10.1016/j.fm.2014.05.009 (Marshall et al. 2014, 149).

[59] Kamal, Rai Aneja, Dhiman Romika, Aggarwal Neeraj Kumar, and Aneja Ashish. 2014. "Review Article Emerging Preservation Techniques for Controlling Spoilage and Pathogenic Microorganisms in Fruit Juices". *Hindawi Publishing Corporation International Journal of Microbiology* 1-14. doi: 10.1155/2014/758942 (Kamal et al. 2014, 5).

[60] Vantarakis, Apostolos, M. Affifi, P. Kokkinos, M. Tsibouxi, and M. Papapetropoulou. 2011. "Occurrence of Microorganisms of Public Health and Spoilage Significance in Fruit Juices Sold in Retail Markets in Greece". *Anaerobe* 17(6):288–91. doi: 10.1016/j.anaerobe.2011.04.005 (Vantarakis et al. 2011, 289).

[61] Bagde, Neha. I., and P. M. Tumane. 2011. "Studies on Microbial Flora of Fruit Juices and Cold Drinks". *Asiatic Journal of Biotechnology Resources* 2(4):454-60. Available online at www.pacificjournals.com (Bagde and Tumane 2011, 458).

[62] Kader Munjur, Abdullah Al Mamun, Tanvir Islam, and Nigarin Sultana. 2014. "Bacteriological Analysis of Some Commercially Packed and Fresh Fruit Juices Available in Jessore City: A Comparative Look". *International Journal of Biosciences* 5(1): 415-420. doi: 10.12692/ijb/5.1.415-420 (Kader et al. 2014, 417-418).

[63] Pylypenko, Inna, Liudmyla Pylypenko, Elena Sevastyanova, Evgeniy Kotlyar, and Ruslana Kruche. 2016. "Epiphytic and Regulated Microbial Contaminants of Food Vegetable Raw Materials and Products". *Food technologies. Ukrainian Food Journal* 5(2):272-80. doi: 10.24263/2304-974X-2016-5-2-6 (Pylypenko et al. 2016, 275).

[64] Adjrah, Y., D. S. Karou, B. Djéri, K. Anani, K. Soncy, Y. Ameyapoh, de C. Souza, and M. Gbeassor. 2011. "Hygienic Quality of Commonly Consumed Vegetables, And Perception About Disinfecting Agents in Lomé." *International Food Research Journal* 18(4):1499-1503. Available at: http://www.ifrj.upm.edu.my/18%20(04)%202011/(41)IFRJ-2011-007.pdf (Adjrah et al. 2011, 1499).

[65] Scientific Committee on Food. 2002. "Risk Profile on the Microbiological Contamination of Fruits and Vegetables Eaten Raw" *European Commission, Health & Consumer Protection Directorate-General, C2 - Management of Scientific Committees II; Scientific Co-Operation and Networks.* 1-45. Available at: https://ec.europa.eu/food/sci-com/scf_en (Scientific Committee on Food 2002, 19).

[66] Okafor, Cordelia N., V. J. Umoh, and M. Galadima. 2003. "Occurrence of Pathogens on Vegetables Harvested from Soil Irrigated with Contaminated Streams". *Science of the Total Environment.* 311: 36 – 49. doi: 10.1016/S0048-9697(03)00057-3 (Okafor et al. 2003, 51).

[67] Waqialla, Samah Mahmoud Ahmed. 2006. *"Microbial Contamination of Fresh Salad Vegetables in Khartoum State Market".* Bsc. Thesis, 1-92. University of Khartoum (Waqialla 2006, 25-26).

[68] Barro, Niolas, A. R. Bello, S. Aly, C. A. T. Ouattara, A. J. Ilboudo and A. S. Traoré. 2006. "Hygienic Status Assessment of Dishwashing Waters, Utensils, Hands, and Pieces of Money from

Street Food Processing Sites in Ouagadougou (Burkina Faso)". *African Journal of Biotechnology* 5(11):1107-12. Available at: http://www.academicjournals.org/AJB (Barro et al. 2006, 1107).

[69] Verhoeff-Bakkenes, Liesbeth. 2012. *"Campylobacter jejuni: Exposure Assessment and Hazard Characterization"*. Thesis. 1-146. Wageningen University, Wageningen, Netherlands (Verhoeff-Bakkenes 2012, 28).

[70] https://vtechworks.lib.vt.edu/bitstream/handle/10919/71797/Alicia%20Thomas%20Final%20Project-pdf.pdf?sequence=1/25.11.2017/6:40.

[71] https://www.health.qld.gov.au/__data/assets/pdf_file/0031/377950/qh-hsdgdl-028-2.pdf.

[72] Berger, Cedric N., Sodha, Samir V., Shaw, Robert K., Griffin, Patricia M., Pink David, Hand Paul, and Frankel Gad. 2010. "Fresh Fruit and Vegetables as Vehicles for the Transmission of Human Pathogens - Minireview". *Environmental Microbiology* 12(9):2385–97. doi: 10.1111/j.1462-2920.2010.02297.x (Berger et al. 2010, 2390).

[73] Hafez, Asghar Ashrafi, Asadolahi, Eghbaleh, Havasian, Mohamadreza, Panahi Jafar, Davoudian, Abdollah, Lotfikar, Mona, and Khosravi, Afra. 2013. "Study on the Parasitic and Microbial Contamination of Vegetables, and the Effect of Washing Procedures on Their Elimination in Ilam City". *Journal of Paramedical Sciences* 4(4):37-41. doi: 10.22037/jps.v4i4.4881 (Hafez et al. 2013, 40).

[74] Issa-Zacharia Abdulsudi, Kamitani Yoshinori, Muhimbula Happiness S., and Ndabikunze Bernadette K. 2010. "A Review of Microbiological Safety of Fruits and Vegetables and the Introduction of Electrolyzed Water as an Alternative to Sodium Hypochlorite Solution". *African Journal of Food Science*. 4(13):778 – 89, http://www.academicjournals.org/ajfs (Issa-Zacharia et al. 2010, 784).

[75] Saccomori, Fernanda, Wigmann, E. F., Bernardi, A. O., Alcano-Gonzalez, M. D. J., Copetti, M. V. 2015. "Influence of storage temperature on growth of *Penicillium polonicum* and *Penicillium*

glabrum and potential for deterioration of frozen chicken nuggets". *International Journal of Food Microbiology* 200:1-4. doi: 10.1016/j.ijfoodmicro.2015.01.014 (Saccomori et al. 2015, 2).

[76] Patrignani, Francesca, Vannini, L., Kamdem, S. L. S., Lanciotti, R., Guerzoni, M. E. 2010. "Potentialities of high-pressure homogenization to inactivate *Zygosaccharomyces bailii* in fruit juices". *Journal of Food Science* 75: 116-20. doi: 10.1111/j.1750-3841.2009.01508.x (Patrignani et al. 2010, 117).

[77] Davidson, P. Michael., and Critzer, F. M. 2012. "Interventions to inhibit or inactivate bacterial pathogens in foods". In *Microbial Food Safety: An Introduction*, edited by Oyarzabal, O. A., Backert, S., 189-202, Springer, New York, (Davidson and Critzer 2012, 194).

[78] Rawat, Seema. 2015. "Food spoilage: microorganisms and their prevention". *Asian Journal of Plant Science & Research* 5(4):47-56. Available online at www.pelagiaresearchlibrary.com (Rawat 2015, 54).

[79] Campos, C. A., Gliemmo, M. F., and Castro, M. P. 2015. "Strategies for controlling the growth of spoilage yeasts in foods". In *Microbial Food Safety and Preservation Techniques*, edited by Rai, V. R., Bai, J. A., 497-511, Taylor & Francis, New York (Campos et al. 2015, 505).

[80] http://www.envirobase.info/PDF/60506EU_REGULATIONS_GUIDE_2006.pdf/09.12.2017/5.30.

In: Microbial Contamination of Food Product ISBN: 978-1-53613-524-4
Editors: G. Mocanu and O. Constantin

Chapter 5

Microbial Contaminants of Alcoholic and Nonalcoholic Beverages

Gabriel – Dănuț Mocanu*
Department of Food Science and Engineering,
Biotechnology and Aquaculture,
Faculty of Food Science and Engineering,
Dunărea de Jos University of Galați,
Galați, România

Abstract

The nonalcoholic and alcoholic beverages presented in this chapter include bottled water, soft drinks, fruit and vegetable juices, wine, and beer, focusing on the microbial contamination of the beverages and the various measures taken to control the spoilage and ensure the products' safety.

Keywords: lactic acid bacteria, yeasts, beverages, contamination

* E-mail: dmocanu@ugal.ro.

INTRODUCTION

There are many fermented alcoholic and nonalcoholic beverages. Two primary products are obtained from fermentations, both used for food preservation (alcohol and acids).

The generation of fermentation products can prevent food spoilage and extend shelf life. Different types of lactic acid bacteria - LAB (e.g., *Lactobacillus* spp., *Leuconostoc* spp., and *Pediococcus* spp.) can grow in wine below anaerobic conditions. Some yeasts and some types of LAB can produce beer spoilage [1].

Raw water can represent a source of spoilage microorganisms for bottled water, because it contains a variety of heterotrophic microorganisms (bacteria, yeasts, and molds) who need organic carbon to grow [2].

The most important microorganisms that can produce the spoilage of soft drinks are yeasts – such as *Saccharomyces* spp., *Pichia* spp., and *Kloeckera* spp. - that form ascospores able to survive thermal treatment and to grow in the final products [3]. Juices are one of the most common and pleasant ways to consume fruits and vegetables. Juices are obtained by pressing fruits or vegetables, and without applying a thermal treatment they are considered to be fresh juices.

The main microorganisms present in juices are yeasts and molds, and sometimes acid-tolerant bacteria. The microbiological stability in juices can be achieved by pasteurization or use of chemical preservatives [4]. In fruit juices the most common microorganisms involved in spoilage are yeasts belonging to the genera *Pichia*, *Candida*, *Hansenula*, *Rhodotorula*, *Saccharomyces*, *Torulopsis*, *Trichosporon* and *Zygosaccharomyces* [5] and molds.

The most dominant molds recorded in fruit juices are *Penicillium* spp.,*Cladosporium* spp.,*Aspergillus niger*, *Aspergillus fumigatus*, *Botrytis* spp., and *Aureobasidium pullulans* [6].

Types and Sources of Spoilage Microorganisms

Bottled Water

Bottled water is usually defined as water with no added ingredients, sealed in bottles or other containers, used for human consumption. The most important source of spoilage microorganisms which can be introduced into bottled water is raw water [7]. Raw water may contain aerobic, Gram negative bacteria belonging to the following genera: *Flavobacterium*, *Acinetobacter*, *Moraxella*, *Pseudomonas*, and *Chromobacterium,* distributed in deep ground water habitats, which have values of 10^6–10^8 CFU/g. In anaerobic aquifer habitats, at levels of 10^5–10^6 CFU/g, the most metabolic groups of bacteria which include nitrate reducers (*Pseudomonas* spp., *Bacillus* spp.), sulfate reducers (*Desulfotomaculum* spp., *Desulfobacter* spp., *Desulfonema* spp.), and iron (Fe(III)) reducers (*Shewanella putrefaciens*) can be found [2].

The water may be contaminated, when it is extracted and collected, by *Bacillus* spp., *Enterobacter* spp., *Klebsiella* spp., *Actinomyces* spp., *Streptomyces* spp., and other microorganisms present on the drilling/pumping equipment and in the soil. The spoilage of bottled water is the result of the regrowth of microorganisms or recontamination after water treatment. These two factors (regrowth and recontamination) are affected by storage temperature, the quantity of nutrients and small levels of disinfectants [6].

Soft Drinks

This category of beverages includes carbonated and noncarbonated products. The absorbtion of carbon dioxide (carbonation) results in carbonated soft drinks which are not alcoholic beverages and are not thermally treated. Noncarbonated drinks do not contain carbon dioxide, are fruit based and are usually pasteurized or chemically preserved to prevent spoilage [8].

The microbiological spoilage of soft drinks usually poses an economic problem. *Zygosaccharomyces* spp., *Brettanomyces* spp., *Saccharomyces* spp., *Candida* spp., *Torulopsis* spp., *Pichia* spp., *Hansenula* spp. and *Rhodotorula* spp.are yeast commonly found in soft drinks. The most important spoilage yeasts present in soft drinks belong to the *Zygosaccharomyces* genus, with *Z. bailii* as the most predominant spoilage yeast [3]. *Z. bailii* can grow in the presence of preservatives and can determine spoilage forming distinct off-odors, off-taste, visible sediment, and increased package pressure. The majority of bacteria in this group will not grow in the high acid medium and the cells will be inactivated. Only a few bacteria are able to grow at a low pH value, among which the most important are *Gluconobacter* spp. and *Acetobacter* spp. In carbonated soft drinks mold spores cannot germinate due to the absence of oxygen and the preservation effect of carbon dioxide. When carbonation is eliminated, due to loss of package integrity, molds can cause spoilage. The most dominant molds recorded in soft drinks are *Aspergillus* spp., *Penicillium* spp., *Rhizopus* spp. and *Fusarium* spp. [9].

Fruit and Vegetable Juice

This category of products is susceptible to spoilage by fermentative yeasts, molds, and a small number of aciduric bacteria. Fruits and vegetables used for juice processing are exposed to microorganism spoilage during production and harvesting, transportation and juice extraction [9]. Fruits and vegetables, during the growing and post-harvesting stages, may be contaminated with a large range of soil-borne yeasts like *Saccharomyces* spp., *Debaryomyces* spp., *Hansenula* spp., *Kluyveromyces* spp., and *Pichia* spp. There are some specific mold species that cause spoilage in fresh fruits and vegetables (e.g., *Byssochlamys* spp., *Talaromyces* spp., *Neosartorya* spp.) [7].

After harvesting, a low number of bacteria, and some spore former bacteria e.g., *Bacillus* spp., *Alicyclobacillus* spp., *Clostridium butyricum*, and *Clostridium pasteurianum* may be found on fresh fruit and vegetables [10].

Fruit juices can exhibit a large number of molds, yeasts, lactic acid bacteria (*Lactobacillus fermentum, Lactobacillus plantarum, Leuconostoc mesenteroides*), and acetic acid bacteria. It has been shown that some pathogens, like the acid-tolerant *Salmonella* and *E. coli* O157:H7 from orange juice and apple cider, can stay viable for a long time (about 30 days) in acid products [11]. Vegetable juices can be contaminated with molds, yeasts, and LAB, along with *Bacillus coagulans*, *Clostridium butyricum*, and *Clostridium pasteurianum* [12].

Fruit and vegetable juices may be affected by some critical intrinsic or extrinsic factors. The category of intrinsic factors includes the juice pH, its oxidation–reduction potential, water activity, and the existence of antimicrobial compounds. The extrinsic factors include packaging material properties, relative humidity, temperatures and storage lengths, and headspace oxygen levels in packages and juice storage tanks.

Wine

Fermented alcoholic beverages are produced by the fermentation of sugars or starches contained in fruits or grains under the action of naturally occurring microorganisms, or by adding starter cultures; through biochemical reactions, these microorganisms change the substrates present in the raw materials into alcohol [13].

Wines are able, under aerobic conditions, to support the growth of film yeasts and acetic acid bacteria (*Acetobacter* spp. and *Gluconobacter* spp.). *Pichia* spp. and *Candida* spp., due to their aerobic nature and fast growth, are capable of forming films, while *Saccharomyces cerevisiae*, *Dekkera bruxellensis*, or *Zygosaccharomyces bailii* can be isolated from wine pellicles [14]. *Saccharomyces cerevisiae* and *Saccharomyces bayanus* are two types of yeast considered desirable in wine fermentation.

Dekkera bruxellensis is another contaminant of wine, due to the production of acetic acid [15]. Under anaerobic conditions, some types of

LAB (e.g., *Lactobacillus*, *Leuconostoc*, and *Pediococcus* spp.) can increase in wine. Other heterofermentative *Lactobacillus* spp. (*Lactobacillus brevis*, *Lactobacillus buchneri*) can ferment glucose and fructose, determining the so-called tourne spoilage defect, expressed through increase in the acidity of the wine. Some types of *Leuconostoc* spp. can determine sliminess and cloudiness. Malic acid can be transformed into lactic acid and CO_2 and reduce the wine acidity with the help of *Oenococcus oeni*. Occasionally, this type of fermentation is used to reduce the sourness of wine [12].

Beer

For centuries, the brewing industry has encountered problems with spoilage bacteria. Some yeasts and some types of LAB (*Lactobacillus brevis*, *Lactobacillus lindneri*) can produce beer spoilage. In this alcoholic beverage the growth of *Pediococcus* spp. produces acidity and cloudiness. Some types of *Lactobacillus* spp. are able to multiply and produce turbidity. In the presence of air, *Acetobacter* spp. and *Gluconobacter* spp. can create cloudiness and sliminess, and cause the sour taste of the final product. Wild yeasts (yeasts other than those normally used in fermentation) can develop in beer, causing off-flavor [12]. Some Gram-negative bacteria such as *Pectinatus cerevisiiphilus*, *Pectinatus frisingensis* and *Megasphaera cerevisiae* are involved in beer spoilage, producing turbidity, acidity and the production of unfavorable smells, such as diacetyl or hydrogen sulphide [16].

Pathogens in Alcoholic and Nonalcoholic Beverages

Pathogenic bacteria can survive in some acidic and carbonated soft drinks, although these microorganisms are not capable of developing during storage. According to Medina et al. [17], some pathogenic bacteria like *E. coli*, *Salmonella enterica*, *Listeria monocytogenes* and

Staphylococcus aureus were not destroyed in Coca Cola after 5 min of incubation. *Yersinia enterocolitica* was inoculated in a commercial orange drink with a pH value of 3.5, and after 3 days at 30 °C it was demonstrated that this pathogenic bacteria was able to survive in the orange drink.

New research about the microbiological quality of carbonated soft drinks developed in Bangladesh showed the following results: in soft drinks, as a result of poor hygiene, the number of pathogenic microorganisms can be very high, and the pathogenic bacteria can stay viable in carbonated soft drinks over long storage periods [18]. The most commonly encountered pathogenic bacteria in fruit juices were *E. coli*, especially the serotype O157:H7, and *Salmonella* (several serotypes) [19]. The most dangerous pathogen related with beverages, due to its reduced infective dose and the seriousness of the disease it causes, is *E. coli* O157:H7 [20].

Until now, in the case of alcoholic beverages, there have been no recorded illness outbreaks determined by pathogenic microorganisms. In regard to these products, only beer contamination has been studied. In the product with 20 EBU bitterness, pH 5.5, and original gravity 1.040, the presence of *E. coli* O157:H7, *Salmonella Typhimurium* and *L. monocytogenes* has been highlighted. These pathogenic bacteria were rapidly inactivated during the fermentation period. With the increase in the pH and the reduction in the concentration of ethanol and original gravity values, the survival of pathogens is possible [21].

CONCLUSION

The major spoilage microorganisms in modern alcoholic and nonalcoholic beverages will probably remain the same as in traditional beverages, but the number of spoilage species is expected to increase. More information is necessary on the behaviour of pathogenic microorganisms in modern beverages to assess the real danger potential. The changes in climate conditions can lead to the emergence of new types of pathogens.

REFERENCES

[1] Suzuki, Koji, S. Asano, K. Iijima, and K. Kitamoto. 2008. "Sake and Beer Spoilage Lactic Acid Bacteria: A Review". *Journal of the Institute of Brewing,* 114(3):209-223. doi: 10.1002/j.2050-0416.2008.tb00331.x. (Suzuki et al. 2008, 209).

[2] Leclerc, Henri, and da Costa, S. Milton. 2005. "Microbiology of natural Mineral Waters". In *Technology of bottled water*, edited by Dorothy Senior and Nicholas Dege. 2nd ed, 325-387 Oxford, UK: Blackwell Publishing Ltd. (Leclerc and da Costa 2005, 357).

[3] Pitt, John I., and A. D. Hocking. 1999. "Fungi and Food Spoilage". In *Yeasts*, 439-468 Gaithersburg, Maryland: Aspen Publishers, Inc. (Pitt and Hocking 1999, 457).

[4] Subramaniam, Persis, and Peter Wareing. 2016. "The Stability and Shelf Life of Food. Second Edition". edited by Woodhead Publishing Series in *Food Science, Technology and Nutrition*, 1-595 Amsterdam (Subramaniam and Wareing 2016, 57, 359).

[5] http://shodhganga.inflibnet.ac.in/bitstream/10603/117765/7/07_chapter%202.pdf/8.12.2017/6.15.

[6] Lawlor, Kathleen A., J. D. Schuman, P. G. Simpson, and P. J. Taormina, 2009. "Microbiological Spoilage of Beverages". In *Compendium of the Microbiological Spoilage of Foods and Beverages*, edited by: W. H. Sperber and M. P. Doyle. 1-367. Springer Science and Business Media, New York, USA (Lawlor et al. 2009, 164, 246-248, 261-267).

[7] Senior, Dorothy, 2005. "Quality management". In *Technology of Bottled Water*, edited by Senior, D. and Dege, N., 272–287. Oxford, UK: Blackwell Publishing Ltd. (Senior 2005, 280).

[8] Fujikawa, Hiroshi, 1997. "Mathematical Models for Thermal Death of Microorganisms". *Journal of Antibacterial and Antifungal Agents, Japan,* 25(9):519-534. (Fujikawa 1997, 527).

[9] ICMSF (International Commission on Microbiological Specifications for Foods), 2011. "Nonalcoholic Beverages". In: *ICMSF. Microorganisms in foods 8: Use of Data for Assessing Process*

Control and Product Acceptance, 269-279. Springer New York Dordrecht Heidelberg London.

[10] ICMSF (International Commission on Microbiological Specifications for Foods), 1998. "Fruits and Fruit Products". In: *ICMSF. Microorganisms in Foods 6: Microbial ecology of food commodities*, 252–273. Gaithersburg, Maryland: Aspen Publishers, Inc.

[11] Raybaudi-Massilia, Rosa M., J. Mosqueda-Melgar, R. Soliva-Fortuny, and O. Martin-Belloso. 2009. "Control of Pathogenic and Spoilage Microorganisms in Fresh-Cut Fruits and Fruit Juices by Traditional and Alternative Natural Antimicrobials". *Comprehensive Reviews in Food Science and Food Safety,* 8(3):157-180. doi: 10.1111/j.1541-4337.2009.00076.x. (Raybaudi-Massilia et al. 2009, 160).

[12] De Bibek, Ray, and A. Bhunia. 2014. *"Fundamental Food Microbiology"*. 5th edition, 41-49, CRC Press Taylor & Francis Group, Boca Raton, (De Bibek and Bhuia 2014, 47).

[13] Jeon, Se Hui, N. H. Kim, M. B., Shim, Y. W. Jeon, J. H. Ahn, S. H. Lee, I. G. Hwang, M. S. Rhee. 2015. "Microbiological Diversity and Prevalence of Spoilage and Pathogenic Bacteria in Commercial Fermented Alcoholic Beverages (Beer, Fruit Wine, Refined Rice Wine, and Yakju)". *Journal of Food Protection,* 78(4):812-818. doi: 10.4315/0362-028X.JFP-14-431. (Jeon and Rhee 2015, 814).

[14] Lambrechts, M. G., I. S. Pretorius. 2000. "Yeast and its Importance to Wine Aroma - A Review". *South African Journal of Enology and Viticulture,* 21:97-129. (Lambrechts and Pretorius 2000, 116).

[15] Renouf, Vincent, and A. Lonvaud-Funel. 2007. "Development of an Enrichment Medium to Detect *Dekkera/Brettanomyces bruxellensis*, a Spoilage Wine Yeast, on the Surface of Grape Berries". *Microbiological Research,* 162(2):154-157. doi: 10.1016/j.micres.2006.02.006. (Renouf and Lonvaud-Funel 2007, 155).

[16] Sakamoto, Kanta, and Koning, N. Wil, 2003. "Beer Spoilage Bacteria and Hop Rezistance – Review Article". *International Journal of Food Microbiology,* 89(2-3):105-124. doi: 10.1016/S0168-1605(03)00153-3. (Sakamoto and Wil 2003, 117).

[17] Medina, Eduardo, Romero, C., Brenes, M. and De Castro, A. 2007. “Antimicrobial Activity of Olive Oil, Vinegar, and Various Beverages Against Foodborne Pathogens”. *Journal of Food Protection,* 70(5):1194–1199. doi: 10.4315/0362-028X-70.5.1194. (Medina and De Castro 2007, 1195).

[18] Akond, Muhammad Ali, S. Alam, S. M. R. Hasan, S. Mubassara, S. N. Uddin, and M. Shirin. 2009. “Bacterial Contaminants in Carbonated Soft Drinks Sold in Bangladesh Markets”. *International Journal of Food Microbiology,* 130(2):156–158. doi: 10.1016/ j.ijfoodmicro.2009.01.014. (Akond et al. 2009, 156-158).

[19] Parish, Mickey, E. 2009. “Food Safety Issues and the Microbiology of Fruit Beverages and Bottled Water”. In: *Microbiologically Safe Foods*, 291–304, edited by N. Heredia, I. Wesley, and S. García, John Wiley & Sons, Inc. (Parish 2009, 300).

[20] Garcia Santos, and N. Heredia, 2009. “Foodborne pathogens and toxins: an overview”. In: *Microbiologically Safe Foods*, 15–52, edited by Heredia, N., I. Wesley, and S. García, John Wiley & Sons, Inc. (Garcia and Heredia 2009, 29).

[21] Menz Garry, Vriesekoop, M. Zarei, B. Zhu, and P. Aldred. 2010. “The Growth and Survival of Food-Borne Pathogens in Sweet and Fermenting Brewers′ Wort”. *International Journal of Food Microbiology,* 140(1):19–25. doi: 10.1016/j.ijfoodmicro.2010.02.01. (Menz et al. 2010, 23).

In: Microbial Contamination of Food Product ISBN: 978-1-53613-524-4
Editors: G. Mocanu and O. Constantin

Chapter 6

MICROBIAL CONTAMINANTS OF CEREALS AND CEREAL PRODUCTS

Corina Neagu and Oana Emilia Constantin*
Department of Food Science and Engineering,
Biotechnology and Aquaculture,
Faculty of Food Science and Engineering,
Dunărea de Jos University of Galați, Galați, România

ABSTRACT

Throughout history, cereals and cereal based products were considered to be the most secure food for human consumption. However, these products may still host a number of hazards that can cause disease to consumers if not managed properly. These hazards are usually microbiological and may occur in the cereals in the field, or during harvesting and storage. The processing of cereal grains in order to obtain flour has a minimal impact on the contamination level of cereals.

Keywords: cereals, bakery products, contamination, mycotoxins, safety

---

* Corresponding Author: econstantin@ugal.ro.

Introduction

According to Alexandros, 2006 [1], cereals continue to remain by far the most important food source in the world, making up 50 per cent of the human calorie intake, and as much as 54 per cent in developing countries. Cereals make up about 55% of the total consumption of protein, 15% of the total consumption of lipids and 70% of the total consumption of carbohydrates; a total of 50-55% of the calories consumed originate from cereals. These foods provide the body with complex carbohydrates, which are an important source of energy, particularly for diets low in fat. Cereals also contain dietary fiber, as well as minerals and vitamins, especially vitamins from the B group, D and E [2].

All around the world, the use of cereals increased by one percent in 2017/18 as compared to the previous year (Figure 6.1). The studies indicate that the consumption of cereals as human food is likely to grow by 1.3%, and as animal feed by 0.4% [3].

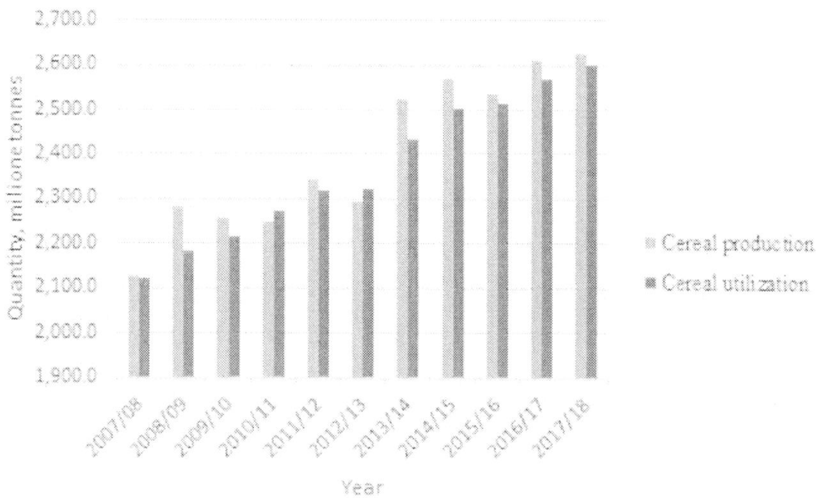

Figure 6.1. Cereal production and utilization worldwide (Adapted from FAO [3]).

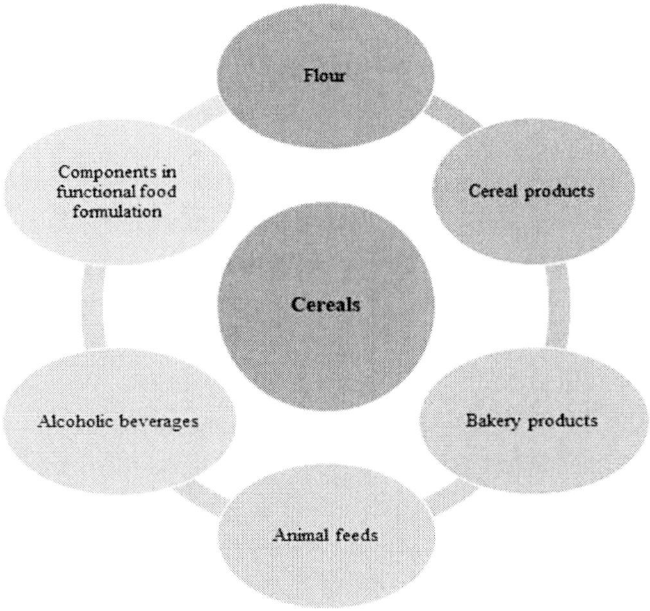

Figure 6.2. The uses of cereals in the world.

The use of cereals worldwide is shown in Figure 6.2.

The food value of a product is given by its nutritional and sensorial values, as well as its degree of safety (absence of pathogenic, toxic substances and spoilage microorganisms), as the microbial contamination is responsible for the decrease of the product's safety and quality.

In nature microorganisms are ubiquitous and they are able to cause food spoilage and/or be responsible for foodborne diseases. Cereals and cereal products can be contaminated with spoilage microorganisms able to render cereal products nonedible (alteration of product taste and the production of undesirable favors), or with pathogenic microorganisms which not only produce changes in the organoleptic characteristics of the products, but that may also cause discomfort and disease [4].

The most important spoilage microorganisms in cereal grains are molds. They cause the heating of grain, off-odors, dry matter loss [4], and their activity causes undesirable changes such as loss of germinability because the molds can consume the seed constituents for their own growth and reproduction [5].

Some species of molds can form allergenic spores and mycotoxins [6]. Mycotoxins in plant products, processed food and feed have a significant economic impact and represent a serious problem for animal and human health [7].

The quality and safety of grain products remain major problems for both food producers and control authorities. Furthermore, food spoilage and the emergence of possible food-caused illnesses can lead to economic losses, due to their negative health effects, productivity reduction, medical expenses, and last but not least the unfavorable advertising for the specialized industry. Food waste is a global problem, therefore in the cereal industry about 5% to 10% of world food supplies are lost annually due to the presence of molds and the by-products of their secondary metabolism (mycotoxins) [4].

Microbial Contamination of Cereals

The microbiota on the surface of cereal grains is varied and it includes bacteria, molds, yeasts and actinomycete. The contamination of cereals occurs on the field and during harvest, transport, and storage. At harvesting time, the populations of fungi and bacteria present in cereal grains can vary between 10^7-10^9 and 10^6-10^8 CFU/g grain [8]. The number and the type of microorganisms present in cereals is influenced by different factors: climate conditions, harvesting equipment, birds, insects and rodent presence, and moisture control during cereal storage [9].

The bacteria which can be found on cereal grains are lactic acid bacteria, rope-forming bacteria (*Bacillus* spp.), and pathogenic bacteria (*Bacillus cereus*, *Clostridium botulinum*, *Clostridium perfringens*, *Escherichia coli*, *Salmonella* spp., and *Staphylococcus aureus)* that can cause safety problems [10]. Coliforms and enterococci also occur as indicators of unsanitary handling and processing conditions [10].

The yeasts present on the cereal grains belong to the genera: *Candida, Cryptococcus, Pichia, Rhodotorula, Sporobolomyces,* and *Trichosporon.* The fungi in the cereal microbiota are classified into two categories: field and storage fungi. Field fungi are present on cereals before harvest and rarely grow during storage; they include *Alternaria alternata*, *Cladosporium herbarum*, *Fusarium culmorum* and *Fusarium graminearum*. Storage fungi are represented by *Aspergillus flavus*, *Aspergillus candidus*, *Penicillium aurantiogriseum*, *Penicillium brevicompactum*, *Eurotium amstelodami* and *Wilemia sebi,* and will grow during storage when environmental conditions allow their development. Storage fungi can contaminate grains during harvest and in contaminated storage facilities [4].

Immediately after harvesting, cereals are not sufficiently dry and can be considered perishable biological products mainly because they may be degraded under the action of microorganisms present on grains from several sources: dust, water, soil, fertilizers, insects, contaminated plants, humans, and animal feces [11]. To prevent the development of molds, it is recommended that after harvesting, the grains be artificially dried, therefore during storage, grain humidity should be below 13%. During the storage period, it is very important to control grain temperature, oxygen concentration and relative humidity, because these factors can favor fungal growth and mycotoxin biosynthesis. If cereals have high humidity during storage it is likely they undergo acid fermentation due to the action of lactic acid bacteria and coliform bacteria, and may be followed by alcoholic fermentation caused by yeasts [4].

The grain quality (dry matter reduction and the ability of a seed to germinate) is affected by poor post-harvest management. The contamination of cereals with molds may modify the carbohydrate and protein composition, with an essential influence on bread and malting quality [8].

Under certain conditions (environmental limits, shortage of food, or competition from other microorganisms), some fungal species that contaminate grains can produce toxic by-products, viz. mycotoxins [6]. Because these substances pose a serious threat to human and animal health, and some of them are very resistant and cannot be eliminated during food processing [12], various research focused on the mitigation strategies of mycotoxins due to their severe human health risks [13-17].

The mycotoxins of major concern for human health are produced by three main genera of fungi: *Aspergillus* produces aflatoxins and ochratoxins A, *Fusarium* - produces fumonisins, zearalenone, and trichothecenes, and *Penicillium* - produces ochratoxins A [18].

Alternaria spp. is able to synthesize numerous secondary metabolites that are classified structurally into five different groups [19-20]:

1. Dibenzo-α-pyrone derivatives: e.g., alternariol (AOH), alternariol mono methylether (AME), altenuene (ALT), isoaltenuene (isoALT) and altenuisol (ATL);
2. Tetramic acid derivatives: e.g., tenuazonic acid (TeA);
3. Perylene quinone derivatives: e.g., altertoxin I, II (ATX-I, -II) and stemphyltoxin III (STTX-III);
4. Aminopentol esters: e.g., AAL TB1 and TB2;
5. Miscellaneous structures: e.g., tentoxin (TEN) and altenuic acid III (AA-III).

These *Alternaria* mycotoxins are known as possible food contaminants with potential toxicological risk and have been found in contaminated agricultural supplies [21].

According to the European Food Safety Authority (EFSA) [20], the AOH and AME toxins are considered as genotoxic substance, TeA was classified as non-genotoxic, but due to its acute oral toxicity in mice and rats, it was classified as possibly harmful, and TEN is considered as non-harmful.

The mechanisms of actions of toxic compounds are summarized in Table 6.1.

Fungal postharvest contamination and the subsequent synthesis of mycotoxins are influenced by different critical factors [18, 43]. These factors are shown in Figure 6.3.

In order to decrease spoilage and also avoid mycotoxin production, without any addition of preservatives, it is necessary to control the environmental factors so as to change the duration of fungal growth.

Table 6.1 Mechanisms of toxicity of main mycotoxins occurring in cereals

Mycotoxin	Mechanisms of toxicity	References
Aflatoxins (AFs)	Pro-inflammatory cytokine IFN-g increasing Anti-inflammatory cytokine IL-4 reduction	[22]
	Interaction with hepatitis B virus in causing hepatocellular carcinoma	
	Interference with nucleotide excision repair	[23]
	Alteration of methylation of genes	
	Change the cytochrome P450 enzymes and glutathione S-transferase activity to toxic hydroxylated metabolites	[24]
Deoxynivalenol (DON)	RNA and DNA synthesis inhibition	[25]
	Tumornecrosis factor alpha (TNF-α)reduction in the plasma	[26]
	Protein biosynthesis inhibition Alteration of the pro-inflammatory cytokine production	[27-28]
Ochratoxins A (OTA)	Generation of reactive oxygen species acting as a genotoxin	[29]
	Causing congenital defects in the fetus	[30]
	Nuclear factor erythroid 2-like inhibition	[31]
Fumonisins (Fum)	Inhibition the enzyme ceramide synthase controlling the acylation of sphinganine	[32]
AOH and AME	Acting as a topoisomerase poison, causing DNA strand breaks.	[33-35]
	In vitro, inhibit the synthesis of progesterone	[36]
	Possess the potential to alter the hormone production and to interfere with gene and receptor expression	[37]
	Causing esophageal cancer	[38-40]
ALT and isoALT	Acting as a topoisomerase poison without affecting DNA activity	[33-35]
ATX-II and STTX-III	Responsible for esophageal cancer. Possess genotoxic potential probably caused by the formation of DNA adducts.	[41]

Adapted from Raiola et al. [42].

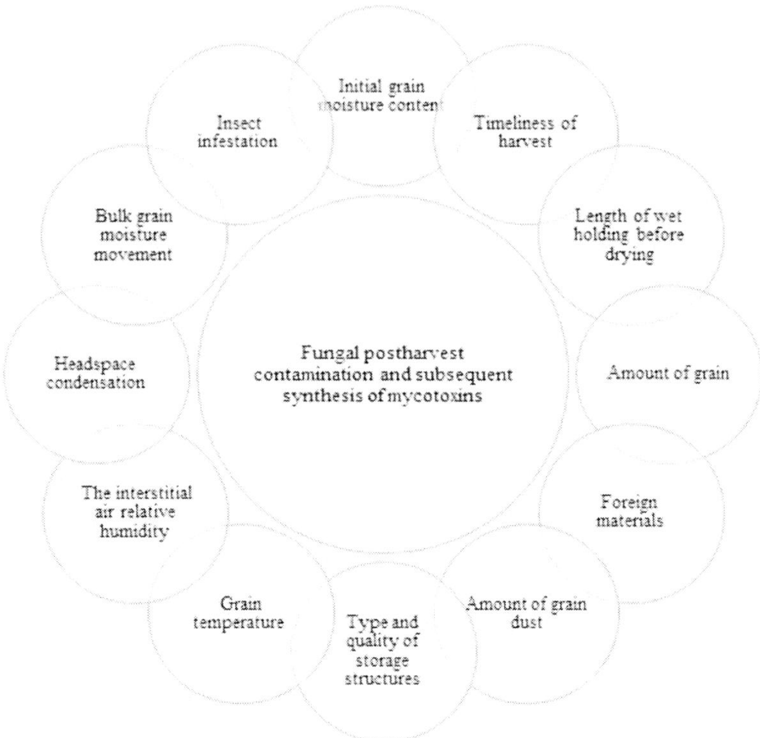

Figure 6.3. Critical factors for fungal postharvest contamination and subsequent synthesis of mycotoxins.

Table 6.2. The minimum water activity requirements of common molds

Group	Species	Minimum a_w
Field fungi	*Alternaria alternata*	0.88
	Cladosporium herbarum	0.88
	Fusarium culmorum	0.89
	Fusarium graminearum	0.89
Storage fungi	*Aspergillus candidus*	0.75
	Aspergillus flavus	0.78
	Eurotium amstelodami	0.71
	Penicillium aurantiogriseum	0.82
	Penicillium brevicompactum	0.80

Source: Adams et al. [45].

Molds usually grow within a temperature range of 10 - 40.5°C, above 70% relative humidity and a pH range of 4 - 8 [44].

The minimum water activity requirements of the most common molds present in cereals is shown in Table 6.2.

Neme and Mohamed [18] classified the factors influencing mycotoxin biosynthesis in grain into five categories: biological factors, environmental factors, harvesting factors, storage factors, handling and processing factors (Figure 6.4).

Mycotoxin biosynthesis in stored grains is possible under the following conditions: the grains have a high moisture degree (16 - 30%), the grain temperature is between 25 - 32°C, and the relative air humidity is high (80 - 100%) [46, 18].

The optimum temperature and water activity for mycotoxin biosynthesis in grains differs according to the types of mycotoxin (Table 6.3) [18-19].

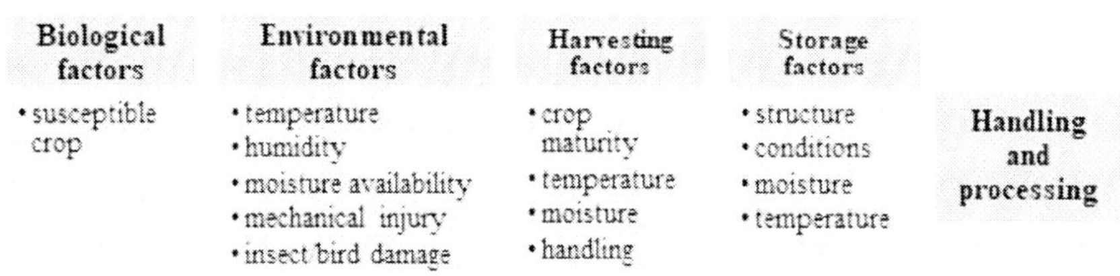

Figure 6.4. Factors affecting mycotoxin contamination of grain.

Table 6.3. Optimum temperature and water activity for mycotoxin biosynthesis

Mycotoxin	Temperature, °C	Water activity
Aflatoxin	33	0.99
Citrinin	20 - 30	0.75 - 0.85
Deoxynivalenol	26 - 30	0.995
Fumonisin	15 - 30	0.9 - 0.995
Ochratoxin	25 - 30	0.98
Zearalenone	25	0.96
AOH, AME, TeA	25 - 30	0.98

Adapted from Neme and Mohammed [18].

Table 6.4. The mycotoxins most commonly associated with cereal grains following pre- and postharvest contamination

Cereals	Pre-harvest	Post-harvest
Barley	DON, NIV, ZEN, HT-2, T-2	OTA, Afs, Cit
Maize	DON, Fum, ZEN	ZEN, Afs
Oats	DON, NIV, HT-2, T-2	OTA, Cit
Sorghum	Ergot	Afs
Wheat	DON, NIV, ZEN	OTA, Afs, Cit

NIV-Nivalenol; ZEN –Zearalenone; Cit- Citrinin; HT-2 – HT-2 toxin; T-2 – T-2 toxin.
Source: Neme and Mohammed [18].

The pH and hydrogen ion concentration of the substrate can influence the interactions between a_w and temperature by the effects generated on metabolic processes (e.g., sporulation and morphogenesis) [47].

The insect infestation of grain is another factor that can determine fungal and mycotoxin contamination [48-49]. During storage, insects, due to their metabolic heat, can increase the water activity and temperature in the grain to levels suitable for fungal growth. Insects generate water by metabolic heat through condensation on surfaces because of temperature variations, which may determine rapid spoilage [50].

The mycotoxins most commonly associated with cereal grains following pre- or postharvest contamination are shown in Table 6.4 [18].

The timely grain harvesting, proper drying and good storage conditions are the basic preventive mechanisms for mold growth and mycotoxin production. Sanitation, screening, aeration and monitoring of stored grain are basic good management practices during storage.

Microbial Contamination of Flour

Flour is an important component in people nutrition on a global level. Flour is obtained from cereals or other starchy food sources by dry milling. It is most commonly made of wheat, but also maize, rye, barley and rice, among many other grasses and non-grain plants. The dry milling process involves three general operations: cleaning, tempering and milling, and these operations cause a redistribution of contaminants. Flour has a lower microbiological load than wheat and the fractions obtained during milling (germ and bran fractions) [51].

After cleaning, the microbial load is reduced to 1 log CFU/g for aerobic bacteria and yeast and mold [52], and also mycotoxin concentration is reduced [53-55].

The milling process does not cause a reduction in the number of microorganisms and does not destroy mycotoxins, but instead it ensures their distribution and concentration in the milled fractions, especially in germs and bran [53, 55-57]. The germ and bran fractions are less used in bakery products; however, the consumption of whole grain and whole grain milled products is on the increase, due to these products' health benefits [58]. In the flour fractions the mycotoxin content is lower than in germs or bran fractions; however, the reduction of mycotoxin content depends on the wheat variety, the penetration capacity of molds in cereals, and the presence of mycotoxins in the internal layer of the seed [59].

The concentration of mycotoxins in cleaned wheat, as compared to that in unclean grains, ranged from 7 to 63% for DON, from 7 to almost 100% for NIV, and from 7 to 40% for ZEA [54, 60-61].

The final microbiological quality of wheat flour can be greatly influenced by these residues if they are not properly controlled [51].

Studies conducted in different countries indicate that the number of yeasts and molds in flour is within the range of 1.3 log and 5.9 log CFU/g [52, 56-57, 62-64].

The molds isolated from flour belong to the following genera: *Penicillium, Aspergillus, Rhizopus, Eurotium, Fusarium, Cladosporium, Mucor, Rhizopus, Alternaria* [65].

Mold spores present in flour can survive for several years, therefore special attention should be given to flour storage. 1 or 2% changes in the moisture content during the storage period may be sufficient for mold growth and mycotoxin production. Furthermore, in flours stored at room temperature with a moisture content of 18% fungal counts at the beginning of storage will arise from 2×10^2 to 2×10^6 CFU/g [64].

In addition to fungi, a wide variety of spoilage bacteria are also present in flour.

A mean aerobic bacterial count reported by different studies is comprised between 1.0 to 7.2 log CFU/g [52, 56-57, 62, 66-67]. Several microbiological surveys on flour indicated the presence of *Bacillus* spp., *Pseudomonas* spp., *Streptococcus* spp., *Achromobacter* spp., *Flavobacterium* spp., *Micrococcus* spp., *Alcaligenes spp., Escherichia coli* and *Salmonella* spp.) [51].

Wheat flour can contain *Bacillus* spp. spores as a consequence of processing or post processing contamination [68]. *Bacillus* spp. is the spoilage contaminant of flour-containing foods (e.g., rope spoilage of bread); however, it is also *associated* with food safety incidents [68-72]. The presence of pathogenic *Bacillus* species, such as *B. cereus*, in flour-containing foods like wheat-pasta is frequently the cause of several foodborne illnesses [69, 73-74].

Although *Salmonella* spp. will not grow in flour and flour meal due to their low water activity (a_w = 0.70), the pathogens can remain viable for months in the raw product. The flour meal is not usually considered an ingredient at risk of contamination with *Salmonella,* because the products are thermally treated before consumption; however, the flour is considered risky when used as raw ingredient for pharmaceuticals, spices or flavorings, and ready-to-eat products, or as a bulking agent in dehydrated mixes [75].

If flour is stored for 12 - 48 months, a reduction of 10 - 100-fold of bacteria and fungi is observed, particularly at temperatures between 10 - 25°C. Usually, the water activity in flour is low, but when the water contents rises these conditions favor microbial growth, especially the germination of *Bacillus* spp., molds of the *Aspergillus, Penicillium* and *Eurotium* genera, and amylolytic yeasts *(Saccharomycopsis* spp., and *Pichia burtonii)* [8].

Microbial Contamination of Bakery Products

Bread is a staple all over the world, accounting for an average of approx. 30% of the daily diet. Bread consumption (particularly white bread) does not provide the nutrients required by the organism, but it still represents an important source of energy provided by its high carbohydrate content. Due to the protein content (approx. 7%), lipids (less than 1%), and carbohydrates (approx. 50%), bread and bakery products are the basis of the food pyramid; the bread also has an important role to play in providing the B-group vitamin content [76].

The characteristics of bakery product technology depend on the raw materials used (e.g., type of flour), additives and manufacturing technology and, furthermore, the eating habits in different places around the world. The bread and bakery products can have positive effects on the population health which is influenced by the variety of these products (breakfast cereal, pasta, bakery and fresh pastries, etc.).

Bakery products are obtained by mixing flour, yeast and salt with water and occasionally with various other ingredients (sweeteners, dairy products, shortenings, eggs, enzymes, preservatives, fibers, fats, solids, etc.). After leavening, the dough obtained by mixing the ingredients is baked, thus giving rise to the desired product.

Several methods can be used to classify bakery products: according to product type, or on the basis of their pH and water activity [77-78]. The classification of bakery products according to product type is shown in Figure 6.5.

Smith et al. 2004 [77] classified bakery products on the basis of pH into three groups:

- *High acid* bakery products with pH < 4.6
- *Low acid* bakery products with pH > 4.6 but < 7
- *Non-acid or alkaline* bakery products with pH > 7

According to the classification based on a_w, bakery products can be categorized as follows [78]:

- *Low-moisture* bakery products ($a_w < 0.6$)
- *Intermediate-moisture* bakery products (a_w between 0.6 and 0.85)
- *High-moisture* bakery products ($a_w > 0.85$ and generally between 0.95 and 0.99).

Bakery products are usually evaluated by consumers according to their sensorial characteristics (odor, taste and appearance). These properties can be influenced by the occurrence of spoilage which can be classified according to [77] into three categories:

- *physical* spoilage (moisture loss and staling),
- *chemical* spoilage (rancidity)
- *microbiological* spoilage (yeast, mold and bacterial growth).

The classification of products according to their pH and a_w can contribute to spoilage identification in bakery products.

Figure 6.5. The classification of bakery products by product type.

Because dough baking is achieved at high temperatures which are responsible for destroying the vegetative forms of bacteria, yeasts and molds, post process contamination can cause the spoilage of bakery products. These types of products are contaminated from different sources such as air, equipment, cooling and slicing, and packaging operations.

The factors that influence the spoilage of bakery products are shown in Figure 6.6.

Microbiological spoilage is the major factor that influences the shelf life of bakery products with a high or intermediate moisture, and it is also a main cause of economic loss in the bakery industry [77]. The spoilage of bakery products caused by molds, yeasts, and rope bacteria is triggered by the following factors: the pH (5.4 - 7.5), the a_w (0.75 - 0.98) and the ambient temperatures. Water activity is the most important factor that results in the spoilage of bakery products.

Figure 6.6. Factors which influenced the spoilage of bakery products.

For low moisture baked products with aw <0.6 microbiological spoilage does not pose a serious problem in the bakery industry. For intermediate moisture products (a_w 0.6 - 0.85), the microorganisms involved in spoilage are osmophilic yeasts and molds, whereas for high moisture content products (a_w 0.94 - 0.99), microbiological spoilage is a major problem because the water activity values may favor the growth of all bacteria and fungi [77].

Bacterial Spoilage

Bacteria need higher values of water activity to develop, and therefore bacterial spoilage is limited to high moisture bakery products. The major bacterial spoilage in bread, doughnuts, crumpets, and cakes is "rope," caused by *Bacillus subtilis*, a spore forming bacteria. This bacterium is usually present in raw ingredients (flour and sugar), and can survive the baking process due to its spores' increased resistance to high temperatures. Moreover, rope spoilage may be determined by other species of *Bacillus*, such as *B. licheniformis*, *B. pumilus*, *B. clausii* and *B. firmus* [4].

Another species of *Bacillus* that can be found in bakery products is at risk of causing illness is *B. cereus*. The ingredients that may be sources of *Bacillus* spp. include dairy products (cream, dried milk [79, 80] and whey concentrate [81], dried eggs [82], yeast and improvers [83-85], soy protein [86], dried fruits [87], cocoa [85], spices, such as ginger, mace, allspice, cinnamon, garlic, and pizza spice [88].

Ropness occurs mostly in high temperature seasons when the climatic conditions, a_w and pH of the baked product favor spore germination. This type of spoilage can be easily identified by consumers because it is characterized by a typical, unpleasant, fruity odor (overripe melons or pineapple) and crumb modification by enzymatic degradation (soft, adhesive, and faded). Enzymatic degradation is determined by the amylase and protease activity, and the adhesiveness by the extracellular polysaccharides produced by the *Bacillus* spp. [4].

Bakery products can also be contaminated after baking by spores of *Bacillus* spp. due to inadequately cleaned and sanitized mixers or bowls during dough production, pipelines, filters and water containers, as well as packaging materials [71]. Since consumption of ropy bread has been associated with food-borne illness [89], it is necessary to take measures to prevent and control rope spoilage. These measures include implementation of good hygiene practices and good manufacturing practices in the factory, use of analysis certificates for external raw ingredients, and also using acid preservatives such as propionic acid and calcium propionate to ensure a low rope spore count [4].

The spores of *B. cereus* can survive the mild heat processing conditions of baking although its thermal resistance is strongly influenced by the pH of the product [90].

Because up to 14% of *B. cereus* strains may be psychrotrophic, custards and fillings need be prepared in small lots, cooled rapidly, and stored at 4°C [90].

The contamination of bakery products with *Staphylococcus aureus* usually occurs because of poor manufacturing practices, and even the ingredients may also be sources of contamination (milk from mastitic cows, reconstituted dried milk, whey, and cream) [91]. Even if *S. aureus* is destroyed by heating, the enterotoxin produced by the microorganism is heat resistant and cannot be inactivated by pasteurization [91].

The bakery products that may contain pasteurized whole egg, egg white, and egg yolk raw eggs, are susceptible of contamination with *Salmonella* spp. Baking or cooking should destroy any *Salmonella* spp. present in the product, but the degree of inactivation will depend on the composition of the bakery product (pH, a_w, preservatives used), level of contamination, and time-temperature control of the heat process for food safety [77]. Wyatt and Guy, 1981, reported that keeping a pumpkin pie product with a pH 5.6 and an a_w of 0.97 at a temperature of 108°C, for 1 minute, was enough to destroy 10 CFU/g of inoculated *S. typhimurium* [92].

Bakery products containing dairy ingredients may be contaminated with *Listeria monocytogenes*. This microorganism can grow at refrigerated storage temperatures on products such as cream, cheese or butter based iced or filled pastries, and cakes that are consumed without heating. Because *L. monocytogenes* is a ubiquitous microorganism especially in the bakery environment, a post processing contamination of end products is possible [77].

To reduce the cross contamination of products by pathogenic bacteria and to limit the spread of this foodborne pathogen, it is necessary to implement a Hazard Analysis Critical Control Point (HACCP) and a continuous training in food hygiene and safety of bakery workers and food handlers [77].

Yeast Spoilage

Yeast spoilage is less common in the bakery industry; the growth of the yeasts causes the surface defects of the products [93].

Yeasts can cause “visible spoilage” manifested by the visible growth of yeast on the surface of the products, and the “fermentative spoilage” of baked products, manifested by alcoholic, esteric, or other odors and/or visible gas production, such as gas bubbles in products with jams and fondant fillings.

Visible spoilage is usually common in the products with high values of water activity, whereas fermentative yeasts can be found in products with a low a_w. *Pichia burtonii* (which causes the “chalk mold” of bread), *Candida guilliermondii*, *Hansenula anomala* and *Debaryomyces hansenii* are examples of yeasts encountered occasionally in bakery products [94].

The surface of bakery products with a high sugar content (40 - 70%) can develop yeasts that are able to tolerate the conditions of low water activity and high osmotic stress (osmotolerance). Typical osmotolerant yeast species belong to the *Debaryomyces*, *Pichia* and *Zygosaccharomyces* genera [95].

Spoilage yeasts are able to convert sugars present in filling creams to ethanol, carbon dioxide, alcohols, glycerol, esters, aldehydes, ketones, sulphur volatiles and organic acids [93]. These final compounds can be responsible for the changes in the acidity and flavor profile of the product, with acetic acid having the strongest sensorial impact [93].

Moreover, members of the *Brettanomyces*, *Candida*, *Kloeckera*, *Rhodotorula, Saccharomyces*, *Schizosaccharomyces* and *Torulopsis* genera are notable as spoilage agents when foods are not produced in accordance with good manufacturing practices.

Mold Spoilage

Molds have a greater contribution than other spoilage microorganisms to the spoilage of bakery products. Spoilage caused by mold results in losses of 1 and 5% of the product. Such losses are influenced by the product type, processing method and climate condition [96]. The shelf life of the bakery products is around 5 - 6 days when they are not preserved. After baking, the bakery products do not contain any viable vegetative molds and mold spores. Mold spoilage results from post baking contamination by spores in the air, surfaces and processing equipment, and raw additives added to the end products [97-99]. If the products are packaged prior to being entirely cooled, condensation may occur inside the package, which modifies the water activity of the product's surface and thus may determine mold growth [79]. The typical genera of molds involved in bakery spoilage are: *Penicillium*, *Aspergillus*, *Cladosporium*, *Fusarium*, *Monilia*, *Endomyces*, *Rhizopus*, *Eurotium*, and *Mucor* [4].

The first phase in mold spoilage is the presence of white, filamentous colonies, which progressively turn into various colors from blue-green to black, as spores are produced [4].

CONCLUSION

Cereal grains and cereal products can provide the perfect environment for spoilage and pathogenic microorganisms, and the control of these microorganisms is very important to reducing contamination, alteration and foodborne diseases. The implementation of good hygiene practices in the plant, the storage of grains and cereal products in controlled atmosphere, the reduction of the value of a_w (e.g., using chemical antimicrobial agents, gas packaging), the use of smart labels can contribute to keeping contamination and microbial growth under control.

References

[1] Alexandratos, Nikos, Jelle Bruinsma, G. Bödeker, J. Schmidhuber, S. Broca, P. Shetty, and M. G. Ottaviani. 2006. *"World Agriculture: Towards 2030/2050. Interim Report. Prospects for Food, Nutrition, Agriculture and Major Commodity Groups."* 1-78. http:// www.fao.org/fileadmin/user_upload/esag/docs/Interim_report_AT20 50web.pdf, Accessed September 12, 2017 (Alexandratos et al. 2006, 52).

[2] Dewettinck, Koen, Filip Van Bockstaele, Bianka Kühne, Davy Van de Walle, T. M., Courtens, and Xavier Gellynck. 2008. "Nutritional Value of Bread: Influence of Processing, Food Interaction and Consumer Perception." *Journal of Cereal Science* 48(2):243-257. doi: 10.1016/j.jcs.2008.01.003 (Dewettinck et al. 2008, 244-245).

[3] Food and Agriculture Organization of the United Nations (FAO). 2017. *"World Food Situation."* http://www.fao.org/ worldfoodsituation/csdb/en/_Accessed September 12, 2017. (Food and Agriculture Organization of the United Nations, 2017).

[4] Wenjian, Yang, Dapeng Li, and Alfred Mugambi Mariga. 2017. "Spoilage Microorganisms in Cereal Products." In *Food Spoilage Microorganisms Ecology and Control*, edited by Yanbo Wang, Wangang Zhang and Linglin, 3-21. CRC Press Taylor & Francis Group, New York (Wenjian et al. 2017, 3-7, 9).

[5] Gupta, Ruchi, and Sheela Srivastava. 2014. "Antifungal Effect of Antimicrobial Peptides (AMPs LR14) Derived from *Lactobacillus plantarum* Strain LR/14 and Their Applications in Prevention of Grain Spoilage." *Food microbiology* 42:1-7. doi: 10.1016/j.fm.2014.02.005 (Gupta and Srivastava 2014, 1).

[6] Benbrook, Charles. 2005. "Breaking the Mold Impacts of Organic and Conventional Farming Systems on Mycotoxins in Food and Livestock Feed." *Organic Center State of Science Review,* 1-58. https://www.organic-center.org/breaking-the-mold-impacts-of-organic-and-conventional-farming-systems-on-mycotoxins-in-food-and-livestock-feed, Accessed October 22, 2017 (Benbrook 2005, 1).

[7] IARC. 1993. "Some Naturally Occurring Substances: Food Items and Constituents, Heterocyclic Aromatic Amines and Mycotoxins." In *IARC Monographs on the Evaluation of Carcinogenic Risks to Humans*, 1-609. Lyon: International Agency for Research on Cancer (IARC 1993, 397-444; 445-466; 467-488).

[8] Magan Nares, and David Aldred. 2006. "Managing Microbial Spoilage in Cereal and Baking Products." In *Food Spoilage Microorganisms*, edited by Clive de W. Blackburn, 194-212. Woodhead Publishing Limited, Cambridge, England (Magan and Aldred 2006, 194-195).

[9] Doyle, Michael P., and Robert L., Buchanan. 2013. "Food Microbiology." In *Food Microbiology: Fundamentals and Frontiers*, edited by Michael P. Doyle and Robert L. Buchanan, 1-1118. ASM Press, Washington, DC (Dole and Buchanan 2013, 213).

[10] Bullerman Lloyd B., and Andreia Bianchini. 2011. "The Microbiology of Cereals and Cereal Products." *Food quality and safety*, February 24. Accessed November 10, 2017. http://www.foodqualityandsafety.com/article/the-microbiology-of-cereals-and-cereal-products, Accessed November 1, 2017 (Bullerman and Bianchini 2011).

[11] Butscher, Denis, Tina Schlup, Christian Roth, Nadina Müller-Fischer, Corinne Gantenbein-Demarchi, and Philipp Rudolf von Rohr. 2015. "Inactivation of Microorganisms on Granular Materials: Reduction of *Bacillus amyloliquefaciens* Endospores on Wheat Grains in a Low Pressure Plasma Circulating Fluidized Bed Reactor." *Journal of Food Engineering* 159:48-56 (Butscher et al. 2015, 48).

[12] Hill, R., and J., Lacey. 1983. "Factors Determining the Microflora of Stored Barley Grain." *Annals of Applied Biology* 102:467-483. doi: 10.1111/j.1744-7348.1983.tb02718.x (Hill and Lacey 1983, 481).

[13] Bullerman, Lloyd B., and Andreia Bianchini. 2014. "Good Food-Processing Techniques: Stability of Mycotoxins in Processed Maize-Based Foods." In *Mycotoxin Reduction in Grain Chains*, edited by Leslie John F., and Antonio Logrieco, 1-376. John Wiley & Sons, Inc. Published (Bullerman and Bianchini 2014, 89).

[14] Didier Jans, Katia Pedrosa, Dian Schatzmayr, Gerard Bertin, and Bertrand Grenier. 2014. "Mycotoxin Reduction in Animal Diets." In *Mycotoxin Reduction in Grain Chains*, edited by Leslie John F., and Antonio Logrieco, 1-376. John Wiley & Sons, Inc. Published (Didier et al. 2014, 101).

[15] Munkvold, Gary P., 2014. "Crop Management Practices to Minimize the Risk of Mycotoxins Contamination in Temperate-Zone Maize." In *Mycotoxin Reduction in Grain Chains*, edited by John F. Leslie, Antonio F. Logrieco, 59-75 (Munkvold 2014, 59).

[16] Ochieng, Peter Juma, Daniel Okun, Samuel Runo, N. J. Njagi, and Jeffry Murage. 2013. "Public Health Strategies for Preventing Aflatoxin Exposure." *British Journal of Cancer* 45:1-22. (Ochieng et al. 2013, 2).

[17] Wild, Christopher P., David J., Miller, and John D., Groopman. 2015. *Mycotoxin Control in Low- and Middle Income Countries*. Lyon IARC Working Group Report (Wild et al. 2015, 9).

[18] Neme, Kumera, and Ali Mohammed. 2017. "Mycotoxin Occurrence in Grains and the Role of Postharvest Management as a Mitigation Strategies. A Review." *Food Control* 78:412-425. doi: 10.1016/j.foodcont.2017.03.012 (Neme and Mohamed 2017, 412, 414, 419).

[19] Zwickel, Theresa, Sandra M. Kahl, Horst Klaffke, Michael Rychlik and Marina E. H. Müller. 2016. "Spotlight on the Underdogs - An Analysis of Underrepresented *Alternaria* Mycotoxins Formed Depending on Varying Substrate, Time and Temperature Conditions." *Toxins* 8:1-17. doi: 10.3390/toxins8110344 (Zwickel et al, 2016, 2).

[20] EFSA. 2011. "Scientific Opinion On the Risks for Animal and Public Health Related to The Presence of *Alternaria* Toxins in Feed and Food." *EFSA Journal* 9(10):2407 (EFSA 2011, 1, 2).

[21] Bruce, Verneal R., Michael E. Stack, and Philip B. Mislivec. 1984. "Incidence of Toxic *Alternaria* Species in Small Grains from the USA." *Journal of Food Science* 49(6):1626-1627 (Bruce et al. 1984, 1626).

[22] Qian, Guoqing, Lili Tang, Xia Guo, Franklin Wang, Michael E. Massey, Jianjia Su, Tai L. Guo, Jonathan H. Williams, Timothy D. Phillips, and Jia-Sheng Wang. 2014. "Aflatoxin B1 Modulates the Expression of Phenotypic Markers and Cytokines by Splenic Lymphocytes of Male F344 Rats." *Journal of Applied Toxicology* 34(3):241-249. doi: 10.1002/jat.2866 (Qian et al. 2014, 241).

[23] Kew, C. Michael. 2013. "Aflatoxins as a Cause of Hepatocellular Carcinoma." *Journal of Gastrointestinal & Liver Diseases* 22:305-310. http://www.jgld.ro/2013/3/13.html Accessed November 11, 2017 (Kew 2013, 305).

[24] Mehrzad, Jalil, Bert Devriendt, Kim Baert, Eric Cox, 2014. "Aflatoxin B1 Interferes with the Antigen-Presenting Capacity of Porcine Dendritic Cells." *Toxicology in vitro* 28:531-537. doi: 10.1016/j.tiv.2013.11.015 (Mehrzad et al. 2014, 531).

[25] Kiessling Karl-Heinz. 1986 "Biochemical Mechanism of Action of Trichothecene Mycotoxins." *Pure & Applied Chemistry* 58(2):327-338. http://old.iupac.org/publications/pac/1986/pdf/5802x0327.pdf Accessed December 11, 2017 (Kiessling 1986, 328-332).

[26] Awad, Wageha, Khaled Ghareeb, Josef Bohm, Jurgen Zentek. 2013. "The Toxicological Impacts of the *Fusarium* Mycotoxin, Deoxynivalenol, in Poultry Flocks with Special Reference to Immunotoxicity." *Toxins* 5(5):912-925. Accessed October 28, 2017. doi: 10.3390/toxins5050912 (Awad et al. 2013, 914).

[27] Desjardins, Anne E. 2006. "Mechanism of Action of Trichothecenes." In: *Fusarium Mycotoxins: Chemistry Genetics and Biology,* edited by Anne E. Desjardins, APS Press, St. Paul, USA (Desjardins 2006, 53-54).

[28] Girgis, George N., John R. Barta, Brash, M., Smith, T. K., 2010. "Morphological Changes in the Intestine of Broiler Breeder Pullets Fed Diets Naturally Contaminated with *Fusarium* Mycotoxins with or without Coccidial Challenge." *Avian Diseases* 54:67-73 doi: 10.1637/8945-052809-Reg.1 (Girgis et al. 2010, 67).

[29] Walker, Ron and John Christian Larsen. 2005. "Ochratoxin A: previous risk assessments and issues arising." *Food additives and*

contaminants 22(s1):6-9. https://www.ncbi.nlm.nih.gov/pubmed/16332615 Accessed November 6, 2017 (Walker and Larsen 2005, 7).

[30] Wangikar, Pralhad B., Prabakar Dwivedi, Anil K. Sharma, Neeraj Sinha and Avinash G. Telang 2005. "Teratogenic effects in rabbits of simultaneous exposure to ochratoxin A and aflatoxin B1 with special reference to microscopic effects." *Toxicology* 215:37-47 (Wangikar et al. 2005, 38).

[31] Jennings, Paul, Christina Weiland, Alice Limonciel, Katarzyna M. Bloch, Robert Radford, Lydia Aschauer, Tara McMorrow T., Anja Wilmes, Walter Pfaller, Hans J. Ahr, Craig Slattery, Edward A. Lock, Michael P. Ryan and Heidrun Ellinger-Ziegelbauer 2012. "Transcriptomic Alterations Induced by Ochratoxin A in Rat and Human Renal Proximal Tubular *In Vitro* Models and Comparison to A Rat *In Vivo* Model." *Archives of toxicology* 86(4):571-589. doi: 10.1007/s00204-011-0780-4 (Jennings et al. 2012, 572).

[32] Solfrizzo, M., S. N., Chulze, C., Mallmann, A., Visconti, A., De Girolamo, F., Rojo, and A., Torres. 2004. "Comparison of Urinary Sphingolipids in Human Populations with High and Low Maize Consumption as a Possible Biomarker of Fumonisin Dietary Exposure." *Food additives and contaminants* 21:1090-1095. doi: 10.1080/02652030400013318. (Solfrizzo et al. 2004, 1091).

[33] Marko, Doris, Gudrun Pahlke, Markus Fehr. 2007. "Alternariol Acts as a Topoisomerase Poison." *Toxicology Letters* 172S:S50. doi: 10.1016/j.toxlet.2007.05.157 (Marko et al. 2007).

[34] Fehr, Markus, Gudrun Pahlke, Jessica Fritz, Morten O. Christensen, Fritz Boege, Martina Altemöller, Joachim Podlech, and Doris Marko. 2009. "Alternariol Acts as a Topoisomerase Poison, Preferentially Affecting the IIα Isoform." *Molecular Nutrition & Food Research* 53(4):441-451. doi: 10.1002/mnfr.200700379 (Fehr et al. 2009, 441).

[35] Pfeiffer, Erika; S., Eschbach., Manfred, Metzler. 2007. "Alternaria Toxins: DNA Strand-Breaking Activity in Mammalian Cells *in vitro*." *Mycotoxin Research* 23(3):152–157. doi: 10.1007/BF02951512 (Pfeiffer et al. 2007, 153).

[36] Tiemann, U., W. Tomek, F. Schneider, M. Mueller, R. Poehland, and J. Vanselow. 2009. "The Mycotoxins Alternariol and Alternariol Methyl Ether Negatively Affect Progesterone Synthesis in Porcine Granulosa Cells *in vitro*." *Toxicology Letters* 186:139-145. doi: 10.1016/j.toxlet.2009.01.014 (Tiemann et al. 2009, 140).

[37] Frizzell, Caroline, Doreen Ndossi, Shewit Kalayou, Gunnar S. Eriksen, Steven Verhaegen, Morten Sørlie, Christopher T. Elliott, Erik Ropstad, and Lisa Connolly. 2013. "An *in vitro* investigation of Endocrine Disrupting Effects of the Mycotoxin Alternariol." *Toxicology and Applied Pharmacology* 271(1):64-71. doi: 10.1016/j.taap.2013.05.002 (Frizzell et al., 2013, 64).

[38] Graf, Eva, Markus Schmidt-Heydt, and Rolf Geisen. 2012. "HOG MAP Kinase Regulation of Alternariol Biosynthesis in *Alternaria alternata* is Important for Substrate Colonization." *International Journal of Food Microbiology* 157(3):353-359. doi: 10.1016/j.ijfoodmicro.2012.06.004 (Graf et al. 2012, 353).

[39] Brugger, Eva-Maria, Jörg Wagner, David M. Schumacher, Karen Koch, Joachim Podlech, Manfred Metzler, and Leane Lehmann. 2006. "Mutagenicity of the Mycotoxin Alternariol in Cultured Mammalian Cells." *Toxicology Letters* 164(3):221-230. doi: 10.1016/j.toxlet.2006.01.001 (Brugger et al. 2006, 221).

[40] Liu, G. T., Y. Z. Qian, P. Z. M. Zhang, Z. Y. Dong, Y. Z. Shi, Zhen, J. Miao, Y. M. Xu. 1991. "Relationships between *Alternaria alternata* and Oesophageal Cancer." *IARC Scientific Publication 105*: 258–262. https://www.ncbi.nlm.nih.gov/pubmed/1855864 Accessed November 11, 2017 (Liu et al. 1991, 258).

[41] Fleck, Stefanie C., Britta Burkhardt, Erika Pfeiffer, and Manfred Metzler. 2012. "*Alternaria* Toxins: Altertoxin II Is a Much Stronger Mutagen and DNA Strand Breaking Mycotoxin Than Alternariol and Its Methyl Ether in Cultured Mammalian Cells." *Toxicology letters* 214(1):27-32. doi: 10.1016/j.toxlet.2012.08.003 (Fleck et al. 2012, 27-28).

[42] Raiola, Assunta, Gian Carlo Tenore, Lara Manyes, Giuseppe Meca, and Alberto Ritieni. 2015. "Risk Analysis of Main Mycotoxins

Occurring in Food for Children: An Overview." *Food and Chemical Toxicology* 84:169-180. doi: 10.1016/j.fct.2015.08.023 (Raiola et al. 2015, 171).

[43] Channaiah, Lakshmikantha H., and Dirk E., Maier. 2014. "Best Stored Maize Management Practices for the Prevention of Mycotoxin Contamination." In *Mycotoxin Reduction in Grain Chains*, edited by John F. Leslie, 78-88, John Wiley & Sons, Inc. Published (Channaiah and Dirk 2014, 78).

[44] Channaiah, Lakshmikantha 2011. "Mycotoxins." *World-Grain.com.* June 14. Accessed October 2, 2017 http://www.world-grain.com/News/News%20Home/Features/2011/6/Mycotoxins.aspx?cck=1 (Channaiah, 2011).

[45] Adams Martin R., Maurice O. Moss, and Peter McClure. 2016 "Microbiology of Primary Food Commodities." In *Food Microbiology*, edited by Martin R Adams, Maurice O Moss, and Peter McClure 130-171, The Royal Society of Chemistry, Cambridge (Adams et al. 2016, 162).

[46] Shanahan, J., W. Jr. Brown, and T. Blunt, 2003. "Aflatoxins." In *Micotoxin Reduction in Grain*, edited by John F. Leslie, Antonio Logrieco, Colorado State University, Cooperative Extension Publication. Crop Series: Production. No. 0.306 (Shanahan et al. 2003, 81-82).

[47] Magan, Naresh, and J. Lacey. 1984. "Effect of Temperature and pH on Water Relations of Field and Storage Fungi." *Transactions of the British Mycological Society* 82:71-81 (Magan and Lacey 1984, 71).

[48] Abbas, Hamed K., Robert M. Zablotowicz, Mark A. Weaver, W. Thomas Shier, H. Arnold Bruns, Nacer Bellaloui, Cesare Accinelli, and Craig A. Abel. 2013. "Implications of Bt Traits on Mycotoxin Contamination in Maize: Overview and Recent Experimental Results in Southern United States." *Journal of Agricultural and Food Chemistry* 61(48):11759-11770. doi: 10.1021/jf400754g (Abbas et al. 2013, 11760).

[49] Jouany, J. P., A. Yiannikouris, and G. Bertin. 2009. "Risk Assessment of Mycotoxins in Ruminants and Ruminant Products."

Options Mediterranéennes, A 85: 205-224. doi: 10.3390/ toxins7083057 (Jouany et al. 2009, 208).

[50] Magan, Naresh, Russell Hope, Victoria Cairns, and David Aldred. 2003. "Post-Harvest Fungal Ecology: Impact of Fungal Growth and Mycotoxin Accumulation in Stored Grain." *European Journal of Plant Pathology* 109(7):723-730. doi: 10.1023/A:1026082425177 (Magan et al. 2003, 728).

[51] Sabillón Galeas, Luis E. 2014. *"Understanding the Factors Affecting Microbiological Quality of Wheat Milled Products: from Wheat Fields to Milling Operations."* Thesis, University of Nebraska (Sabillón Galeas 2014).

[52] Manthey, Frank A., Charlene E., Wolf-Hall, Saujanya Yalla, Chitra Vijayakumar, and Darla Carlson. 2004. "Microbial Loads, Mycotoxins, and Quality of Durum Wheat from the 2001 Harvest of the Northern Plains Region of the United States." *Journal of food protection* 67(4):772-780 (Mathey et al. 2004, 777).

[53] Abbas Hamed K., C. J. Mirocha, R. J. Pawlosky and D. J. Pusch. 1985. Effect of Cleaning, Milling, and Baking on Deoxynivalenol in Wheat. *Applied and Environmental Microbiology* 50(2):482-486. doi: 10.4315/0362-028X-67.4.77 (Abbas et al. 1985, 484).

[54] Lancova, Katerina, Jana Hajslova, Marta Kostelanska, J. Kohoutkova, Jan Nedelnik, Hana Moravcova, and Marie Vanova. 2008. "Fate of Trichothecene Mycotoxins During the Processing: Milling and Baking." *Food additives and contaminants* 25(5):650-659. doi: 10.1080/02652030701660536 (Lancova et al. 2008, 652).

[55] Pascale, Michelangelo, Miriam Haidukowski, Veronica Maria Teresa Lattanzio, Marco Silvestri, Roberto Ranieri, and Angelo Visconti. 2011. "Distribution of T-2 and HT-2 Toxins in Milling Fractions of Durum Wheat." *Journal of Food Protection* 74(10):1700-1707. doi: 10.4315/0362-028X.JFP-11-149 (Pascale et al. 2011, 1703).

[56] Richter K. S., E., Dorneanu, K. M., Eskridge, and C. Rao. 1993. *Microbiological Quality of Flours. Cereal Foods World.* 38:367-369 (Richter et al. 1993, 367).

[57] Berghofer, Lana K., Ailsa D. Hocking, Di Miskelly, and Edward Jansson. 2003. "Microbiology of Wheat and Flour Milling in Australia." *International Journal of Food Microbiology* 85(1):137-149. doi: 10.1016/S0168-1605(02)00507-X (Berghofer et al. 2003, 138).

[58] Sloan, Elizabeth A. 2011. "Consumer Trends: Consumers go with the grain." *Food Technology - Chicago* 65(12):35-42. http://scitechconnect.elsevier.com/wp-content/uploads/2015/12/Consumer-Trends-in-Grain-Consumption-1.pdf Accessed October 28, 2017 (Sloan 2001, 37).

[59] Pinson-Gadais, Laetitia, Christian Barreau, Marc Chaurand, Stephanie Gregoire, Magalie Monmarson, and Florence Richard-Forget. 2007. "Distribution of Toxigenic *Fusarium* spp. and Mycotoxin Production in Milling Fractions of Durum Wheat." *Food Additives and Contaminants* 24(1):53-62. doi: 10.1080/02652030600892958 (Pinson et al. 2007, 58-58).

[60] Edwards, Simon G., E.T. Dickin, Susan MacDonald, D. Buttler, C.M. Hazel, S. Patel, and K.A. Scudamore. 2011. "Distribution of *Fusarium* Mycotoxins in UK Wheat Mill Fractions." *Food Additives & Contaminants: Part A* 28(12):1694-1704. doi: 10.1080/19440049.2011.605770 (Edwards et al. 2011, 1698).

[61] Neuhof, Torsten, Matthias Koch, Tatjana Rasenko, and Irene Nehls. 2008. "Occurrence of Zearalenone in Wheat Kernels infected with *Fusarium culmorum*." *World Mycotoxin Journal* 1(4):429-435. doi: 10.3920/WMJ2008.1055 (Neuhof et al. 2008, 432).

[62] Seiler, D. A. 1986. "The Microbial Content of Wheat and Flour." In *Spoilage and Mycotoxins of Cereals and Other Stored Products: Proceedings of The Spring Meeting of the Biodeterioration Society* 22:35-40 (Seiler 1986, 37-38).

[63] Batool, Syeda A., Naseem Rauf, S.S., Tahir, and Razia Kalsoom. 2012. "Microbial and Physico-Chemical Contamination in the Wheat Flour of the Twin Cities of Pakistan." *International Journal of Food Safety* 14:75-82. http://www.internetjfs.org/articles/Microbial and

Physico-chemical contamination in the wheat flour of the twin cities of Pakistan.pdf Accessed October 28, 2017 (Batool et al.2012, 79).

[64] Weidenbörner, Martin, C. Wieczorek, S. Appel, and B. Kunz. 2000. "Whole Wheat and White Wheat Flour - the Mycobiota and Potential Mycotoxins." *Food Microbiology* 17:103–107. doi: 10.1006/fmic.1999.0279 (Weidenbörner et al. 2000, 106).

[65] Plavšić, Dragana V., Marija M. Škrinjar, Đorđe B. Psodorov, Ljubiša Ć. Šarić, Dragan Đ. Psodorov, Ana O. Varga, and Anamarija I. Mandić. 2017. "Mycopopulations of Grain and Flour of Wheat, Corn and Buckwheat." *Food and Feed Research* 44(1):39-45. doi: 10.5937/FFR1701039P (Plavšić et al. 2017, 40).

[66] Eglezos, Sofroni. 2010. "Microbiological Quality of Wheat Grain and Flour from Two Mills in Queensland, Australia." *Journal of Food Protection* 73(8): 1533-1536. https://www.ncbi.nlm.nih.gov/pubmed/20819368 Accessed October 6, 2017 (Eglezos 2010, 1534).

[67] Sperber, William H., and The North American Millers' Association Microbiology Working Group. 2007. "Role of Microbiological Guidelines in the Production and Commercial Use of Milled Cereal Grains: A Practical Approach for the 21st Century." *Journal of Food Protection* 70:1041-1053. doi: *10.4315/0362-028X-70.4.1041* (Sperber and The North American Millers' Association Microbiology Working Group 2007, 1042).

[68] Sorokulova, Irina B., O. N. Reva, V. V. Smirnov, Irina V. Pinchuk, S. V. Lapa, Maria C. Urdaci. 2003. "Genetic Diversity and Involvement in Bread Spoilage of *Bacillus* Strains Isolated from Flour and Ropy Bread." *Letters in Applied Microbiology* 37:169-173. doi: 10.1046/j.1472-765X.2003.01372.x (Sorokulova et al. 2003, 169).

[69] Dierick, Katelijne, Els Van Coillie, Izabela Swiecicka, Geert Meyfroidt, Hugo Devlieger, Agnes Meulemans, Guy Hoedemaekers, Ludo Fourie, Marc Heyndrickx, and Jacques Mahillon. 2005. "Fatal Family Outbreak of *Bacillus Cereus* - Associated Food Poisoning." *Journal of Clinical Microbiology* 43(8):4277-4279. doi: 10.1128/JCM.43.8.4277-4279.2005 (Dierick et al. 2005, 4277).

[70] Fangio, Maria Florencia, Sara Ines Roura, and Rosalia Fritz. 2010. "Isolation and Identification of *Bacillus* spp. and Related Genera from Different Starchy Foods." *Journal of Food Science* 75: M218-M221. doi: 10.1111/j.1750-3841.2010.01566.x (Fangio et al. 2010, M218).

[71] Thompson, Jacquelyn M., Will M., Waites, and Christine E. R. Dodd. 1998. "Detection of Rope Spoilage in Bread Caused by *Bacillus* species." *Journal of Applied Microbiology* 85:481-486. doi: 10.1046/j.1365-2672.1998.853512.x (Thompson et al. 1998, 481).

[72] Bennett, Reginald W., and Negash Belay. 2001. "*Bacillus cereus*." In *Microbiological Examination of Foods*, edited by P. F. Downes, and K. Ito, 311-324. Washington: American Public Health Association, Washington, DC (Bennett and Belay 2001, 312).

[73] Delbrassinne, Laurence, Mirjana Andjelkovic, Andreja Rajkovic, Nadine Bottledoorn, Jacques Mahillon, and Joris Van Loco. 2011. "Follow-up of the *Bacillus cereus* Emetic Toxin Production in Penne Pasta under Household Conditions using Liquid Chromatography Coupled with Mass Spectrometry." *Food microbiology* 28(5):1105-1109. doi: 10.1016/j.fm.2011.02.014 (Delbrassinne et al. 2011, 1105).

[74] Naranjo, María, Sarah Denayer, Nadine Botteldoorn, Laurence Delbrassinne, Jean Veys, Jacques Waegenaere, Nicolas Sirtaine, Ronald B., Driesen, Karin R., Sipido, Jacques, Mahaillon and Katelijne, Dierick. 2011. "Sudden Death of a Young Adult Associated with *Bacillus cereus* Food Poisoning." *Journal of Clinical Microbiology* 49(12):4379-4381. doi: 10.1128/JCM.05129-11 (Naranjo et al. 2011, 4379).

[75] Akins-Lewenthal, Deann. 2014 "Flour and Meal." In *The Microbiological Safety of Low Water Activity Foods and Spices* edited by Joshua B. Gurtler, Michael P. Doyle, and Jeffrey L. Kornacki, 245-267. Springer, New York (Akins-Lewenthal 2014, 246-247).

[76] Le Bail, Alain, Constantina Tzia, and Virginia Giannou. 2006. "Quality and Safety of Frozen Bakery Products." In *Handbook of*

Frozen Food Processing and Packaging, edited by Da-Wen Sun, 502-528. Taylor & Francis (Le Bail et al. 2006, 503).

[77] Smith, James P., Daphne Phillips Daifas, Wassim El-Khoury, John Koukoutsis, and Anis El-Khoury. 2004. "Shelf Life and Safety Concerns of Bakery Products - A Review." *Critical Reviews in Food Science and Nutrition* 44(1):19-55. doi: 10.1080/10408690490263774 (Smith et al. 2004, 20, 22, 47, 49).

[78] Smith, James P., and B. K. Simpson. 1995. "Modifed Atmosphere Packaging of Bakery and Pasta Products." In *Principles of Modifed-Atmosphere and Sous Vide Product Packaging*, edited by Jeffrey M. Farber and Karen Dodds, 207-215. Technomic Publications, Lancaster, P.A (Smith and Simpson 1995, 207).

[79] Larsen, H Daugaard and Jorgensen, K. 1997." The Occurrence of *Bacillus cereus* in Danish Pasteurized Milk." *International Journal of Food Microbiology* 34(2):179-186. doi: 10.1016/S0168-1605(96)01182-8 (Larsen and Jorgensen 1997, 179).

[80] Slaghuis, Betsie A., Meike. C. Te Giffel, Rijkelt R. Beumer and Geert Andre. 1997. "Effect of Pasturing on the Incidence of *Bacillus cereus* Spores in Raw Milk." *International Dairy Journal* 7:201–205. doi: 10.1016/S0958-6946(97)00012-5 (Slaghuis et al. 1997, 201).

[81] Pirttijarvi, Tuija S., L. M. Ahonen, Liisa M. Maunuksela, and Marja S. Salkinoja-Salonen. 1998. "*Bacillus cereus* in a Whey Process." *International Journal of Food Microbiology* 44(1):31-41. doi: 10.1016/S0168-1605(98)00117-2 (Pirttijarvi et al. 1998, 31).

[82] Shafi, Rehana, O. J. Cotterill, and Mary L. Nichols. 1970. "Microbial Flora of Commercially Pasteurized Egg Products." *Poultry Science* 49:578-585. doi: 10.3382/ps.0490578 (Shafi et al. 1970, 578).

[83] Collins, Nicola E., Lee Ann M. Kirschner, and Alexander von Holy. 1991. "Characterization of *Bacillus* Isolates from Ropey Bread, Bakery Equipment and Raw Materials." *South African Journal of Science* 87:62-66 (Collins et al. 1991, 62).

[84] Bailey, C. P., and Alexander von Holy. 1993. "*Bacillus* Spore Contamination Associated with Commercial Bread Manufacture." *Food Microbiology* 10(4): 287-294 (Bailey and von Holy1993, 287).

[85] Te Giffel, C. Meike, R. Rijkelt, Beumer, S. Leijendekkers, and M. Frank Rombouts. 1996. "Incidence of *Bacillus cereus* and *Bacillus subtilis* in Foods in the Netherlands." *Food Microbiology* 13:53-58 (Te Giffel et al. 1996, 53).

[86] Becker, Heinz, Gabriele, Schaller, Wolfgang von Wise, and Gerhard Terplan. 1994. "*Bacillus cereus* in Infant Foods and Dried Milk Products." *International Journal Food Microbiology* 23(1):1-15 (Becker et al. 1994, 1).

[87] Aidoo, Kofi E., Richard F. Tester, Janet E. Morrison, and David MacFarlane. 1996. "The Composition and Microbial Quality of Pre-Packed Dates Purchased in Greater Glasgow." *International Journal of Food Science & Technology* 31(5):433-438. doi: 10.1046/j.1365-2621.1996.00360.x (Aidoo et al. 1996, 433).

[88] Kneifel, Wolfgang, and Erika Berger. 1994. "Microbiological Criteria of Random Samples of Spices and Herbs Retailed on the Austrian market." *Journal Food Protection* 57(10):893-901. doi: 10.4315/0362-028X-57.10.893 (Kenifel and Berger 1994, 893).

[89] Yibar, Artun, Figen Çetinkaya, and Gul E., Soyutemiz. 2012. "Detection of Rope-Producing *Bacillus* in Bread and Identification of Isolates to Species Level by Vitek 2 System." *Journal of Biological and Environmental Sciences* 6(18):243-248. http://agris.fao.org/agris-search/search.do?recordID=TR2016009566 Accessed December 8, 2017. (Yibar et al. 2012, 244).

[90] Bassen, M. K., Lokesh K. Gupta, L. Jolly, and Ram P. Tewari. 1989. "Thermal Resistance of *Bacillus cereus* Spores in Custard Preparations." *World Journal of Microbiology and Biotechnology* 5(4):511-516. doi: 10.1007/BF01741827 (Bassen et al. 1989, 511).

[91] Bergdoll, M. S. 1989. "*Staphylococcus aureus*." In *Foodborne Bacterial Pathogens*, edited by. M. P. Doyle, and Marcel Dekker, 463 - 523. New York: Marcel Dekker, Inc., New York (Bergdoll 1989, 463).

[92] Wyatt, Jane C., and V. H. Guy. 1981. "Growth of *Salmonella typhimurium* and *Staphylococcus aureus* in Retail Pumpkin Pies."

Journal of Food Protection 44(6):418-421. doi: 10.4315/0362-028X-44.6.418 (Wyatt and Guy 1981, 418).

[93] Fleet, Graham. 1992. "Spoilage Yeasts." *Critical Reviews in Biotechnology* 12(1-2):1-44 doi: 10.3109/07388559209069186 (Fleet 1992, 1).

[94] Legan, David J., and Philip A. Voysey. 1991. "Yeast Spoilage of Bakery Products and Ingredients." *Journal Applied Bacteriology* 70(5):361-371. doi: 10.1111/j.1365-2672.1991.tb02950.x (Legan and Voysey 1991, 361).

[95] Bubnová, Michala, Jana Zemančíková, and Hana Sychrová. 2014. "Osmotolerant Yeast Species Differ in Basic Physiological Parameters and in Tolerance of Non-Osmotic Stresses." *Yeast* 31(8):309-321. doi: 10.1002/yea.3024 *(*Bubnova et al. 2014, 309).

[96] Guynot, Maria, Antonio J. Ramos, Sala, D., Vicent Sanchis, and Sonia Marın. 2002. "Combined Effects of Weak Acid Preservatives, pH and Water Activity on Growth of *Eurotium* Species on a Sponge Cake." *International Journal of Food Microbiology* 76:39-46. doi: 10.1016/S0168-1605(01)00751-6 (Guynot et al. 2002, 39).

[97] Jarvis, B. 1972. "Mold Spoilage of Foods." *Process Biochemistry 7:11*-14 (Jarvis 1972, 11).

[98] Seiler, D. A. L. 1978. "The Microbiology of Cake and Its Ingredients." *Food Trade Review* 48(6):339-344 (Seiler 1978, 339).

[99] Seiler, D. A. L. 1988. "Microbiological Problems Associated with Cereal Based Foods." *Food Science Technology Today* 2:37-40 (Seiler 1988, 37).

In: Microbial Contamination of Food Product ISBN: 978-1-53613-524-4
Editors: G. Mocanu and O. Constantin

Chapter 7

MICROBIAL CONTAMINANTS OF ACIDIFIED SPECIALTY PRODUCTS

Gabriel-Dănuț Mocanu **[*]** ***and Elisabeta Botez***
Department of Food Science, Food Engineering,
Biotechnology and Aquaculture,
Faculty of Food Science and Engineering,
Dunărea de Jos University of Galați, Galați, România

ABSTRACT

This chapter summarizes the degradation by a variety of microorganisms of acidified specialty food products like mayonnaise, dressings, condiment sauces, mustard and vinegar. These foods are the most microbiologically stable and safe food products.

Keywords: water-to-oil emulsion, *Geotrichum* spp., *Aspergillus* spp., *Bacillus coagulans*, *Zygosaccharomyces bailii*, *Lactobacillus fructivorans*

---

[*] Corresponding Author: dmocanu@ugal.ro.

Introduction

The terms sauces and dressings relate to emulsions such as mayonnaise, salad dressing, mustard, vinegar and condiment sauces, such as barbecue sauce, ketchup and spaghetti sauce. All these liquid or semi-solid food products change the sensorial characteristics of the food to which the sauces or dressings have been added [1]. The acidified specialty products, often packaged and distributed without heat treatments in unopened recipients, are microbiologically stable indefinitely at ambient temperatures. These packaged acidified products are frequently intended for multiple applications, exposing them to contamination with microorganisms at the place of consumption in many occasions. The extraordinary resistance to microbiological spoilage of these products is attributable to their pH values, currently below 4.0, the high concentrations of acetic acid, introduced as vinegar, and the limited headspace and oxygen availability.

These conditions prevent the increase rate of most bacteria and all bacterial pathogens. The types of spoilage microorganisms present in acidified specialty products are restricted to different yeasts, molds and bacteria that can tolerate acetic acid at low pH values in the product [2]. The acidified specialty products discussed in this chapter, used in homes or restaurants, are mayonnaise, sauces and salad dressings, ketchup, mustard, and vinegar.

Spoilage Microorganisms

The high acetic acid concentration and low pH in acidified specialty products stop the increase rate of bacterial spore formers and almost all vegetative bacteria, except several species of lactic acid bacteria. These microorganisms contaminate specialty products through air, the ingredients added and the equipment used in preparation. When these products are stored at room temperature, for a long time, most microorganisms present in acidified specialty products die, with the exception of microorganisms capable of growth in an acid environment [3].

Among the microorganisms capable of growth in an acidic environment, the ones that were isolated were various species of *Lactobacillus* (*L. fructivorans*, *L. brevis*, *L. buchneri*, and *L. plantarum*), and *Bacillus* spp. (*B. subtilis*, and *B. mesentericus*), whose numbers did not normally exceed 10/g, molds (*Geotrichum* spp. and *Aspergillus* spp.), yeasts (*Saccharomyces* spp.) [4].

According to Sperber [2] and Sharpe and Pettipher [5] ¾ of the acidified specialty products are damaged by yeasts [2, 5]. Moreover, sixty-two species of yeasts capable of spoiling these products with low pH value were found. Table 7.1 shows seven different yeast species, mostly isolated from spoiled products.

Saccharomyces cerevisiae is the most commonly isolated yeast, an important spoilage microorganism in acidified specialty products. Yeasts are usually aerobic microorganisms, but most can easily increase under anaerobic conditions when there is a fermentable carbohydrate available. Some types of yeasts, like *Cryptococcus* spp. and *Rhodotorula* spp., are strictly aerobic [6]. As thin films or pellicles, these aerobic microorganisms can grow only on the surface of acidified specialty products. In these products, the growth of film yeasts is mainly prevented by the limited availability of oxygen. In the spoilage process of acidified specialty products, molds are involved in exceptional cases, as they do not tolerate the existence of acetic acid. Moreover, molds are obligatorily aerobic with an oxidative metabolism and the condition existing in many high acidic products limits the oxygen content necessary for mold growth. The mold spoilage of these products can take place in the presence of oxygen, for example on the inside of partially emptied recipients and on the outside of recipients that get contaminated with products and are not properly cleaned. The main mold species that can grow in these conditions are *Moniliella acetobutans*, *Penicillium glaucum*, and *Monascus ruber* [7-8].

Table 7.1. Most frequently isolated yeasts from acidified specialty products [6]

Yeast	Frequency (%)
Saccharomyces cerevisiae	7.37
Pichia anomola	5.52
Zygosaccharomyces bailii	5.52
Candida parapsilosis	5.26
Debaryomyces etchellsii	4.73
Issatchenkia orientalis	3.95
Pichia membranaefaciens	3.95

POSSIBLE FACTORS AFFECTING MICROBIAL SPOILAGE

pH and Acidulant Type

The pH values for this category of products range between 3.2-3.9, and the most commonly used acidulant is acetic acid. These pH values effectively prevent the increase of microorganisms, and an important supplementary barrier to those microorganisms that can grow below pH 4.0 is provided by acetic acid. For example, yeasts, in the presence of the most commonly used inorganic and organic food acidulants, can grow at a minimum pH value of 2.0. When the acetic acid is the acidulant, the same types of yeasts cannot grow below a pH value of 2.9.

The strong odor and aroma of acetic acid, added to these products under the form of vinegar, has determined some factories to obtain acidified specialty products with a soft odor and aroma, by using other acidulants instead of vinegar. This attempt should be undertaken with special care and attention because the replacement of acetic acid by a "soft" acidulant can compromise the quality (microbiological safety and stability) of the newly obtained products [9].

Water Activity

A significant influence on the growth of the microorganisms is water activity (a_w), which is the report between the vapor pressure of the food and the vapor pressure of pure water. For most microorganisms, the optimum values of a_w, are within the range 0.995-0.980. The composition of the microflora is influenced by water activity, microorganisms having various requirements with regard to a_w [10].

The values of water activity for most products in this category (acidified specialty products) are very high and allow the growth of most bacteria, yeasts, and molds. For high-fat products like mayonnaise, the water activity has a low value - about 0.90 - which is enough to prevent the increase of most bacteria. The aqueous phase of mayonnaise can be 15–20% lower than the product weight. All the fat-insoluble solutes which are important to aromatize the whole product mass, salts and sugars, are concentrated [2].

Headspace Oxygen Levels

The spoilage caused by molds and oxidative yeast in acidified specialty products is greatly reduced by using acetic acid and by packaging the final products. Molds and oxidative yeasts present in most products in this category can grow only in the presence of molecular oxygen, because they are necessarily aerobic. In addition to the restricted headspace atmosphere, the oxygen is quickly removed in well-sealed packages by the oxidation of the oil present in the product. The repeated adding of atmospheric oxygen may increase spoilage by aerobic fungi in those acidified specialty products that are used on several occasions, once a week or a month [2].

Thermal Treatment

Several acidified specialty products such as mayonnaise are chemically pasteurized by their low values of pH and high content of acetic acid. To provide microbiological stability during shelf life, ketchup – a product with a low acetic acid content, is pasteurized at 85-90°C [11].

Sources of Spoilage Microorganisms

Oil, water, salt, and vinegar - important ingredients used to prepare acidified specialty products - are in general free of microorganisms and cannot cause the spoilage of the final product [11]. The other ingredients, including eggs and liquid egg yolks, dry or liquid sugar, herbs, starches, and spices are more susceptible to microorganism contamination and can contribute to product spoilage. The various sugar ingredients can be the source of osmophilic yeasts (e.g., *Zygosaccharomyces bailii*). The other remaining minor components are known to be sources of molds, yeasts, and lactic acid bacteria. In different types of mustard seeds, the number of lactobacilli varies between $4 \cdot 10^3$ and $4 \cdot 10^5$/g [12]. It is quite likely that the direct source of spoilage microorganisms is the production environment, utensils and processing equipment, as well as the process materials. Cross-contamination from these sources mainly occurs when the processing and packaging equipments are not sufficiently cleaned or sanitized.

In acidified specialty products only a few types of microbiological spoilage defects were observed. Off-flavors may be caused by the increase and fermentation by means of yeasts and heterofermentative lactic acid bacteria. Oxidative yeasts, in the presence of molecular oxygen, can grow as a film or pellicle on the surface of semi-solid or liquid salad dressings. Different types of molds can determine surface discoloration when adequate oxygen is available [2].

TYPES OF ACIDIFIED SPECIALTY PRODUCTS

Mayonnaise

Mayonnaise is a significant ingredient for the production of a number of mayonnaise-based products [13]. In Europe the word mayonnaise is defined as a condiment sauce that is produced by emulsifying the aqueous phase and vegetable oil (52-80%) (usually sunflower, soybean or canola [rapeseed] oil) consisting of acidifying agents such as vinegar and/or lemon juice, egg yolk as emulsifier (at least 6%), and optional ingredients. Besides vinegar and water, the aqueous phase of the mayonnaise may include seasoning additives such as: salts, sugar, various types of condiments and spices, stabilizers, antioxidants, colors, and flavor enhancers (such as monosodium glutamate) [11, 13].

From a microbiological point of view, mayonnaise and related products (such as ketchup, dressing, mustard) can be considered stable, due to their high oil content (70–80%) and existence of acidifying agents. Acidifying agents such as vinegar and/or lemon juice reduce the pH value of the final product and contribute to the desired aroma [13, 14].

Mayonnaise with high pH values (above 4.1) runs a potential risk of food contamination with *Salmonella* spp., *E. coli* O157:H7, and *L. monocytogenes*, because these pathogens are able to survive in this product [15]. The insufficient acidifying agent's concentration or the unpasteurized egg or egg yolk may increase the probability of developing *Salmonella* spp. Most of the reported cases of *Salmonella* spp. refer to homemade mayonnaise [16, 17].

In a study conducted by Jun et al. [16], it was noticed that some types of acidifying agents, such as citric or acetic acid, affected the survival of *Salmonella* spp. In the acetic acid or citric acid solution, the content of *Salmonella* spp. was significantly lower as compared to the one in vinegar and lemon juice. To observe de evolution of *Salmonella enterica* subsp. *enterica* serovar *Typhimurium* at different temperatures and pH values present in commercial mayonnaise, Sun-Kyung et al. [18] conducted a study where they noticed that the inactivation rates were faster at low pH and higher temperature values. The influence of temperature on the death rate of *E. coli* O157:H7 in reduced-calorie mayonnaise dressings was also investigated [19]. The initial viable cell population was 2.23 log CFU/g after 4 days at 30°C, although after 7 days at 20°C, and 58 days at 20°C no viable cells were detected. The authors suggested that the reduction of *E. coli* O157:H7 is very rapid at temperatures at which mayonnaise is stored, distributed, and offered for retail sale.

It has been demonstrated that acetic acid used in commercially produced mayonnaise reduces the pH level and causes the inactivation of the populations of *Salmonella* spp. and *L. monocytogenes*. If the acetic acid concentration is high enough, the risk of (re)contamination with pathogens is reduced [17].

According to Smittle [20], a low level of mayonnaise pH prevents the increase of pathogens like *Clostridium botulinum*, *Cl. perfringens*, *Staphylococcus aureus*, and *Bacillus cereus* in this product. The resistance of *S. aureus* to acid pH for different acidified specialty products was studied by Erdogrul et al. [21]. They observed that the population of *S. aureus* in mayonnaise decreases rapidly when this product was kept at 4 or 25°C, and after 48 h of storage no viable cells of *S. aureus* were recovered from the mayonnaise sample.

Lactobacilli – acid-tolerant microorganisms - can survive and produce undesirable modifications, affecting the sensorial characteristics of the product (appearance, texture, odor and flavor), decrease shelf life and cause economic losses [13]. The most important and frequently encountered microorganism causing the spoilage of mayonnaise and related products belongs to the *Lactobacillus* genus, i.e., *L. fructivorans* [22]. Furthermore, other species of *Lactobacillus* were isolated from acidified specialty products, such as *L. brevis*, *L. casei*, *L. plantarum*, and *L. buchneri*.

Other microorganisms responsible for the spoilage of mayonnaise and dressings are acid-resistant yeasts like *Zygosaccharomyces bailii* and *Pichia membranaefaciens*. These yeasts can grow in the presence of a 3% acetic acid medium. Other types of yeasts associated with the spoilage of acidified specialty products according to Tornai-Lehoczki [23] are

Issatchenkia orientalis, *Candida. inconspicua*, *Candida. parapsilosis*, *Saccharomyces exiguous*, *Saccharomyces dairensis*, *Debaryomyces hansenii*, *Rhodotorula mucilaginosa*, and *Torulopsis cutaneum*.

Molds are rarely responsible for mayonnaise spoilage, because they cannot proliferate in a medium with ≥0.5% acetic acid. The spoilage of mayonnaise and related acidified specialty products according to Sagdic et al. [17] is related to some species of molds like *Geotrichum* spp., *Monascus ruber*, *Moniliella acetoabutans*, and *Penicillium glaucumi*.

Sauces and Salad Dressings

Sauces and salad dressings are used because they change the sensorial properties of foods [17]. Salad dressings have an a_w of 0.92 and a pH value between 3.2-3.9. These acidified specialty products contain 30% or more vegetable oil and 0.9%-1.2% acetic acid [4]. For this product category the commonest microorganisms causing spoilage are *Z. bailii* and *L. fructivorans*. Both microorganisms produce carbon dioxide by the rapid fermentation of fructose. This type of sugar appears in salad dressings as a result of the acid hydrolysis of sucrose.

Contaminations of different types of commercial salad dressings with *Salmonella* spp., *E. coli* O157:H7, and *L. monocytogenes* were investigated by Beuchat et al. [24]. In this research, five separate strain mixtures of each pathogen were used for the inoculation of salad dressings, which were then preserved for 15 days at a temperature of 25°C. It was reported that all pathogens tested died rapidly after 15 days in all the samples.

Ketchup

Ketchup, also called catsup, is considered a shelf-stable product, used as a condiment for various food products. This acidified specialty product is made of tomatoes, sugar, water, vinegar, salt, starch, and spices like onion, garlic, and cinnamon, it has a low pH, i.e., between 3.5-4.0, and an acetic acid content of 0.8–1.0%. Normally, ketchup is made by mixing all ingredients, and preserved by pasteurization at 85–90°C to set the starch, followed by hot filling into bottles [11, 25]. The most common microorganism for tomato-based products is *Bacillus coagulans*; this pathogen is entirely inhibited at low concentrations of acetic acid. Other microorganisms responsible for the spoilage of ketchup containing 0.8% acetic acid are *Saccharomyces* spp. [26].

Mustard

Another highly flavored acidified specialty product, used as a condiment is mustard. This product is prepared by using these ingredients: 17–30% mustard seeds, 1.5–2.5% acetic acid, 2.5–6.5% salt, and 4–8% oil. The final product has a pH value between 3.8–4.3. The microorganisms most likely to growth in mustard are *L. plantarum* and *Z. bailii*. Usually the microbiological contamination of mustard can be prevented by the use of 0.025% sodium bisulfite, or 0.1% sorbic acid or benzoic acid [2].

4-hydroxybenzyl isothiocyanate and allyl isothiocyanate are two antimicrobial compounds produced by mustard seeds. These compounds can prevent de microbiological spoilage of the product [11, 27]. The majority of the microflora present upon examination of mustard comes from spices, and this microflora is destroyed by the mustard ingredients [12]. It was demonstrated that the addition of 0.3-1.5% mustard to mayonnaise destroys the existing microorganisms [28].

Vinegar

The most important ingredient used to produce all the products described is vinegar. This ingredient is manufactured by the oxidation of ethanol to acetic acid by means of *Acetobacter aceti*. *Acetobacter aceti* is a Gram-negative bacterium that grows and forms a pellicle on the surface of juices, forming the “mother” of vinegar. Many molds, yeasts, and bacteria are capable of tolerate vinegar and grow at acetic acid concentrations of 9% (w/w) [2]. Table 7.2 shows different aceto-tolerant spoilage microorganisms that develop in the pH range of 3.5–4.0.

Table 7.2. Aceto-tolerant spoilage microorganisms [2]

Microorganism	Acetic acid (% w/w)
Molds	
Moniliella acetoabutans	9.0
Cladosporium suaveolens	8.0
Monascus ruber	1.8
Penicillium glaucum	1.4
Yeasts	
Saccharomyces acidifaciens	3.5
Pichia membranaefaciens	3.0
Bacteria	
Lactobacillus acetotolerans	5.0
Lactobacillus buchneri	2.5

Conclusion

A few microorganisms can grow in these products because these foods are considered stable and safe. To increase their shelf life (and safety), refrigerated storage is recommended. To verify the growth of spoilage microorganisms, other preservation methods are recommended, such as chemical preservation.

References

[1] Jones, Adrian A. 2000. "Ambient-Stable Sauces and Pickles." *Shelf-life evaluation of foods*, edited by C. M. D. Man, and A. A. Jones, 211-226, Aspen Publishers, Inc., Gaithersburg, Maryland (Jones 2000, 211).

[2] Sperber, William H. 2009. "Microbiological spoilage of acidified specialty products." In *Compendium of the Microbiological Spoilage of Foods and Beverages*, edited by W. H. Sperber, and M. P. Doyle, 285–299, New York: Springer (Sperber 2009, 285, 287, 290, 297).

[3] Smittle, Richard B. 2000. "Microbiological Safety of Mayonnaise, Salad Dressings, and Sauces Produced in the United States: A Review." *Journal of Food Protection* 63(8):1144–1153. doi:0.4315/0362-028X-63.8.1144 (Smittle 2000, 1147).

[4] De Bibek, Ray, and A. Bhunia. 2014. *"Fundamental Food Microbiology,"* Fifth edition," 41 – 49, CRC Press Taylor & Francis Group, Boca Raton, (De Bibek and Bhunia 2014, 47).

[5] Sharpe, M. E., and G. L. Pettipher. 1983. "Food Spoilage by Lactic Acid Bacteria." *Economic Microbiology*. 8:199-223. (Sharpe and Pettipher1983, 199-223).

[6] Deak, Tibor, and L. R. Beuchat. 1996. *"Handbook of Food Spoilage Yeasts,"* 124-125, New York: CRC Press (Deak and Beuchat 1996, 124-125).

[7] Hocking, A. D., 1994. "Fungal Spoilage of High-Fat Foods." *Food Australia*. 46(1):30–33, (Hocking 1994, 30-33).

[8] Pitt, John, I. and A. D. Hocking. 1997. *"Fungi and Food Spoilage (2nd ed.),"* 3-9, Melbourne: Academic & Professional (Pitt and Hocking1997, 6).

[9] Vermeulen, An, T. D. T. Dang, A. H. Geeraerd, K. Bernaerts, J. Debevere, J. Van Impe. 2008. "Modelling the Unexpected Effect of Acetic and Lactic Acid in Combination with pH and a_w on the Growth/no Growth Interface of *Zygosaccharomyces bailii*." *International Journal of Food Microbiology*. 124(1):79–90. doi:10.1016/j.ijfoodmicro.2008.02.020, (Vermeulen et al. 2008, 85).

[10] Kreyenschmidt, J., and R. Ibald. 2012. "Modeling Shelf Life Using Microbial Indicators." *Shelf Life Assessment of Food*, edited by Nicoli, M. C., 127-168, CRC Press, Boca Raton, FL (Kreyenschmidt and Ibald 2012, 153).

[11] Michels, M. J. M., and W. Koning. 2000. "Mayonnaise, Dressings, Mustard, Mayonnaise-based Salads, and Acid Sauces." In *The*

microbiological safety and quality of food, edited by B. M. Lund, T. C. Baird-Parker, & G. W. Gould, 807-835, Aspen Publishers, Gaithersburg, MD, (Michels and Koning 2000, 809).

[12] Kneifel, W., and S. Slama. 1992. "Microflora of Mustard and its Components." *Archiv für Lebensmittelhygiene (Journal of Food Safety and Food Quality)*. 43:141-143, (Kneifel and Slama 1992, 141-143).

[13] Fialová, Jana, J. Chumchalová, K. Mikova, and I. Hrüsova. 2008. "Effect of Food Preservatives on the Growth of Spoilage Lactobacilli Isolated From Mayonnaise-based Sauces." *Food Control*. 19(7):706-713. doi:10.1016/j.foodcont.2007.07.018, (Fialová et al. 2008, 709).

[14] Depree, J. A., and G. P. Savage, 2001. "Physical and Flavour Stability of Mayonnaise." *Trends in Food Science and Technology*. 12(5-6):157-163. doi:10.1016/S0924-2244(01)00079-6, (Depree and Savage 2001, 160).

[15] Erkmen, Osman, and F. T. Bozoglu, 2016. *"Food Microbiology – Principles into Practice,* volume 1-Microorganisms related to foods, food borne disease and food spoilage," 74-80, Wiley, (Erkmen and Bozoglu 2016, 74).

[16] Li, Jun, Li. J. R. Zhu, and J. R. Chen. 2012. "Survival of *Salmonella* in Home-style Mayonnaise and Acid Solutions as Affected by Acidulant Type and Preservatives." *Journal of Food Protection*. 75(3):465-471. doi:10.4315/0362-028X.JFP-11-373, (Zhu et al. 2012, 468).

[17] Sagdic, O., F. Tornuk, S. Karasu, M. Z. Durak, and M. Arici, 2017. "Microbial Ecology of Mayonnaise, Margarine, and Sauces." In *Quantitative Microbiology in Food Processing Modeling the Microbial Ecology*, edited by Anderson de Souza Sant'Ana, 519-532, The Atrium, Southern Gate, Chichester, West Sussex, UK, John Wiley & Sons, Ltd, (Sagdic et al. 2017, 520-522).

[18] Heo, Sun-Kyung, J. Y. Lee, S. H. Jo, D. H. Bae, Y. S. Kim, M. S. Chung, and S. D. Ha. 2010. "Inactivation of *Salmonella Enterica* Serovar *Typhimurium* in Commercial Mayonnaise at Different Temperatures and pH." *Journal of the Korean Society for Applied*

Biological Chemistry. 53(3):351-355. doi:10.3839/jksabc.2010, (Heo et al. 2010, 352).

[19] Hathcox, A. K., L. R. Beuchat, and M. P. Doyle, 1995. "Death of Enterohemorrhagic *Escherichia coli O157:H7* in Real Mayonnaise and Reduced Calorie Mayonnaise Dressing as Influenced by Initial Population and Storage-Temperature." *Applied and Environmental Microbiology*. 61(12):4172-4177. (Hathcox et al. 1995, 4173).

[20] Smittle, Richard, B. 1977. "Microbiology of Mayonnaise and Salad Dressing: a review." *Journal of Food Protection*. 40(6):415-422. doi:10.4315/0362-028X-40.6.415, (Smittle 1977, 420).

[21] Erdogrul, Özlem, F. Erbilir, and S. Toroglu. 2006. "Survival of Acid-Adapted and Non-Adapted *Staphylococcus Aureus* in Various Food Smples." *Annals of Microbiology*. 56:25-27. doi:10.1007/BF03174965, (Erdogrul et al. 2006, 25-27).

[22] Smittle, Richard, B. and R. S. Flowers. 1982. "Acid Tolerant Microorganism Involved in the Spoilage of Salad Dressing." *Journal of Food Protection*. 45(10):977-983. doi:10.4315/0362-028X-45.10.977, (Smittle and Flowers 1982, 979).

[23] Tornai-Lehoczki, Judit, G. Peter, and D. Dlauchy. 2003. "CHROMagar *Candida* Medium as a Practical Tool for the Differentiation and Presumptive Identification of Yeast Species Isolated from Salads." *International Journal of Food Microbiology*. 86(1-2):189-200. doi:10.1016/S0168-1605(03)00256-3, (Tornai-Lehoczki et al. 2003, 195).

[24] Beuchat, Larry, R. J. H. Ryu, B. B. Adler, and M. D. Harrison. 2006. "Death of *Salmonella*, *Escherichia Coli O157:H7* and *Listeria Monocytogenes* in Shelf Stable, Dairy-based, Pourable Salad Dressings." *Journal of Food Protection*. 69(4):801–814. doi:10.4315/0362-028X-69.4.801, (Beuchat et al. 2006, 809).

[25] Bjorkroth, Johanna, K. and H. J. Korkeala. 1997. "*Lactobacillus fructivorans* Spoilage of Tomato Ketchup." *Journal of Food Protection*. 60(3):505-509. doi"10.4315/0362-028X-60.5.505, (Bjorkroth and Korkeala 1997, 507).

[26] Islam, Md. Shahidul, A. Inoue, N. Igura, M. Shimoda, and I. Hayakawa, 2006. "Inactivation of *Bacillus* Spores by the Combination of Moderate Heat and Low Hydrostatic Pressure in Ketchup and Potage." *International Journal of Food Microbiology*. 107(2):124–130. doi:10.1016/j.ijfoodmicro.2005.08.021, (Islam et al. 2006, 128).

[27] Buchta, V., E. Sláviková, R. Vadkartiová, S. Alt, and P. Jilek, 1996. "*Zygosaccharomyces bailii* as a Potential Spoiler of Mustard." *Food Microbiology*. 13(2):133–135. doi:10.1006/fmic.1996.0017, (Buchta et al. 1996, 133-135).

[28] Radford, S. A., and R. G. Board. 1993. "Review: Fate of Pathogens in Home-Made Mayonnaise and Related Products." *Food Microbiology*. 10(4):269–278. doi:10.1006/fmic.1993.1031, (Radford and Board 1993, 274).

About the Editors

Gabriel-Dănuț Mocanu, PhD
Lecturer
Department of Food Science and Engineering,
Biotechnology and Aquaculture,
Faculty of Food Science and Engineering,
Dunărea de Jos University of Galați,
Galați, România
Email: dmocanu@ugal.ro

Dr. Gabriel – Dănuț Mocanu, obtained his BS in Food Technology and Quality Control in 2005 and MSc in Food Control and Expertise in 2007. His PhD degree was received in 2011 from *Dunărea de Jos* University of Galati. He is involved at the undergraduate level in teaching Unit Operations in Food Industry, Power, Machines and Facilities for Holding Agrotourism, Food Packaging for Catering and Agrotourism. His current research areas of interest are: functional foods for disease prevention, dairy processing, and food microbiology, physical and engineering properties of foods, emerging technologies applied for food processing.

Oana Emilia Constantin, PhD
Lecturer
Department of Food Science and Engineering,
Biotechnology and Aquaculture,
Faculty of Food Science and Engineering,
Dunărea de Jos University of Galați,
Galați, România
Email: econstantin@ugal.ro

Dr. Oana Emilia Constantin, received her Bachelor of Science degree in 2005 and PhD degree in 2009 from *Dunărea de Jos* University of Galati. The main areas of interest of the author/editor are: food microbiology and food microbiology control; applied biotechnology: starter cultures, biochemical and structural changes of biological active compounds, environmental biotechnology (waste biovalorisation, waste-water treatment, and bioremediation).

INDEX

I

K

L

M

N

O

P

R

S

T

V

W

Y

Z